Network Governance of Global Religions

Routledge Research in Information Technology and Society

1. Reinventing Government in the Information Age
International Practice in IT-Enabled Public Sector Reform
Edited by Richard Heeks

2. Information Technology in Government
Britain and America
Helen Margetts

3. Information Society Studies
Alistair S. Duff

4. National Electronic Government
Building an Institutional Framework for Joined Up Government—A Comparative Study
Edited by Martin Eifert and Jan Ole Püschel

5. Local Electronic Government
A Comparative Study
Edited by Helmut Drüke

6. National Governments and Control of the Internet
A Digital Challenge
Giampiero Giacomello

7. The Politics of Cyberconflict
Security, Ethnoreligious and Sociopolitical Conflicts
Athina Karatzogianni

8. Internet and Society
Social Theory in the Information Age
Christian Fuchs

9. Hacking Capitalism
The Free and Open Source Software Movement
Johan Söderberg

10. Urban Youth in China: Modernity, the Internet and the Self
Fengshu Liu

11. Network Governance of Global Religions
Jerusalem, Rome, and Mecca
Michel S. Laguerre

Network Governance of Global Religions

Jerusalem, Rome, and Mecca

Michel S. Laguerre

Routledge
Taylor & Francis Group
NEW YORK AND LONDON

First published 2011
by Routledge
605 Third Avenue, New York, NY 10017
4 Park Square, Milton Park, Abingdon, Oxon OX14 4RN

Routledge is an imprint of the Taylor & Francis Group, an informa business

Typeset in Sabon by IBT Global.

First issued in paperback 2013

Library of Congress Cataloging-in-Publication Data

Laguerre, Michel S.
Network governance of global reglions: Jerusalem, Rome, and Mecca / by Michel S. Laguerre.
p. cm. — (Routledge research in information technology and society; 11)
Includes bibliographical references and index.
1. Religions. 2. Communication—Religious aspects. 3. Telecommunication. 4. Polity (Religion) 5. Globalization—Religious aspects. I Title
BL85.L28 2011
201'.6384—dc22
2010041521

ISBN13: 978-0-415-84903-6 (pbk)
ISBN13: 978-0-415-88879-0 (hbk)
ISBN13: 978-0-203-82389-7 (ebk)

In memory of the victims of the devastating earthquake that struck Haiti's capital city, Port-au-Prince, and surrounding communities on January 12, 2010 at 4:53 PM.

Contents

Tables

Preface

The study of network governance, management, and administration of global religions including Judaism, Catholic Christianity, and Islam is in an embryonic state. Since their inceptions, global religions have dealt with issues of geographical expansion, border-crossing communication, and long distance patronage. In the particular case of the Catholic Church, territorial and global governance is set in Canon law that establishes the administrative structure of practices and the hierarchical architecture of authority. In contrast, Judaism and Islam, which have evolved under different matrix forms of governance, have developed separate procedures and rules to regulate relations between religious leaders and between administrative offices within their respective faiths. Each global religion, in practice, operates under a specific form of network governance, which reflects and explains its different mode of organizational management.

Network governance of global religions is not well understood, partly because of a dearth of expert analyses on the issue. Scholarship on religion tends to focus on other aspects of religious life and institutional practices, such as State and Church relations, spirituality, doctrines, rituals, authority, religious architecture, pilgrimages, and other dimensions of the faith. Recent interest in the study of the globalization process has provided an incentive for further explorations of global religions such as sharing features similar to those found in the network governance of multinational or multicountry organizations.

Jerusalem, Rome, and Mecca evoke a vision of symbolic centers and sites of global governance of Judaism, Catholicism, and Islam. No other cities compete with them for such a distinction. In fact, network governance of each global religion is deterritorialized to encompass actors located across the global landscape. In each, one finds sites that carry more weight than others because of their demographic size, wealth, power, structural locations in the network, or simply for reasons of history and tradition. To uncover, delineate, and map linkages between nodes that constitute these networks required the mobilization of a host of friends, colleagues, acquaintances, facilitators, and interviewees. They provided information, advice, or interpretation in the course of the investigation. In addition, three anonymous external reviewers of the manuscript offered useful and suggestive comments for corrections and revision.

The Berkeley Center for Globalization and Information Technology at the Institute of Governmental Studies of the University of California has been the principal site of the cogitation of my ideas on network governance. I particularly want to thank Jack Citrin, Marc Levin, Barbara Campbell, Nick Robinson, and the rest of the staff for their help through the years. The following graduate and undergraduate students—Lauren Kaplan, Kimberly Wolf, Zared Amirkhanian, Michael Chongpornprasert, Manar Hassan, Lauren Holley, Robert Klein, Nasir Mohammed, Carla Federis, Carmen Taylor, Mariam Yousuf, Allison Dolan, Celine Piser, Alina Shlyapochnik, Nicole Anand, Jonathan Bradshaw, Carmel Javier, Ryan Miller, Kate Tiedeman, Teresa Caturano, Steven Chen, Pamela Thornton, Nan Zhang, Jennifer Wu, Meera Garud, Erika Mark, Aileen Ford, Celina Keshishian, Roopika Subramanian, Chinbo Chong, Roohie Ebrahim, Kacie MacGregor, Flavia Garcia, Charles Li, Caitlin Knowlton, and Lia Economos—provided research assistance to the project in terms of bibliographical searches using digital databases available on campus, translations of Arabic texts, transcriptions of interviews, word-processing blocks of citations, proofreading, and checking the alphabetical order and accuracy of references cited. Their everyday presence in the office uplifted my spirit and kept the project going until its completion. For their contributions to the research process, I acknowledge their help and express my gratitude for their collaboration in the production of this piece of scholarship.

I want to express my debt of gratitude to Gabriel Sheffer and Ruth Kark, both at the Hebrew University of Jerusalem, and to Rabbi Naftali Rothenberg at the Van Leer Jerusalem Institute, for their contributions to the segment of the field research in Jerusalem. Others, including Jewish, Muslim, and Catholic religious leaders with whom I consulted both in the United States and abroad in the course of the research, also deserve my special thanks. They shared with me both information and interpretation with the understanding that their identities would not be revealed to protect their privacy. I am nevertheless indebted to every one of them along with the specialists who have labored in this field for many years; I gratefully recognize the influence of their works on my thinking.

I owe a special debt of gratitude to librarians at the Institute of Governmental Studies, the Haas Business School, and the Doe Library at the University of California at Berkeley, the Graduate Theological Union, the Jewish National and University Library, the Hebrew Union College in Jerusalem, the Hebrew University of Jerusalem, and the Bibliothèque Nationale de France in Paris.

I earlier discussed the contents of the chapters in my graduate seminars on "Religion and Globalization" and "Diaspora, Citizenship, and Transnationalism." In the process, I benefited from comments by graduate students and postdoctoral fellows who attended these meetings. I also read a French version of Chapter 1 ("Réseaux de Gouvernance des Religions Mondiales: Problèmatique d'un Nouvel Objet d'Etude") as a keynote address at the 75e Congrès de l'Association Francophone pour le Savoir (ACFAS) in the Section "Citoyenneté, Ethnicité, Diversité: Interrogations Locales, Nationales et Internationales," organized by the Centre de Recherche sur l'Immigration, l'Ethnicité et la Citoyenneté and held on May 11–12, 2007, at the Université du Québec à Trois-Rivières; I am grateful to Micheline Labelle and Jean-Claude Icart for the invitation to participate in this academic meeting. I presented a portion of Chapter 5 in a public lecture titled "Accommodements Raisonnables, Racisme et Politique de la Reconnaissance: Un Regard International," organized by L'Observatoire International sur le Racisme et les Discriminations and sponsored by Le Ministère de l'Immigration et des Communautés Culturelles du Québec, held May 14, 2007, at the Université du Québec à Montréal. I read Chapter 2 at an international symposium on "At Home in Diaspora/Diaspora at Home" organized by the Taube Center for Jewish Studies at Stanford University and held April 25–26, 2010. It benefited from insightful comments by Charlotte E. Fonrobert and other members of the audience. Chapter 4 was presented in the panel on "Global Multistakeholder: Networked Governance for Information and Communication Technology Policy" at the annual meetings of the American Political Science Association, held in Boston in 2008. I am grateful to the discussant, Lee McKnight as well as to Sam Cherribi and Hatem Bazian.

Last, but not least, I want to acknowledge with gratitude the commissioning editor for Routledge Research in Information Technology and Society Series, Max Novick, who smoothly maneuvered the manuscript through the peer-review and editorial process, the quality control supervisor, Melissa Andros, who oversaw the production of the book, and graduate research assistant Amy Lee, who diligently prepared the Index. The proceeds from the royalty of the sale of the book will be donated to the Catholic Parish of Lascahobas in care of the poor of this middle-sized town, located on the Central Plateau of Haiti.

The focus on network governance in the study of religion expresses the view or idea that an analysis of its organizational infrastructure can yield a productive outcome and a new way to circumvent global phenomena. Global religion is unveiled as a common faith with its local, regional, national, and transnational specificity. A network theory of exchange, reciprocity, and solidarity is also established explaining why a node or a set of nodes may come to rescue a vulnerable unit to ensure the viability of the network. In this light, the deconstruction of a local practice helps us understand unambiguously a dimension of the global.

Rome, Italy
June 2009

Introduction

This sociological study, carried out under the auspices of the Berkeley Center for Globalization and Information Technology at the Institute of Governmental Studies of the University of California at Berkeley, seeks to explain three models of network governance embedded in digital practices that the mainstream monotheistic religions—Judaism, Catholic Christianity, and Islam—have used to lead and manage the worldwide distribution of their local nodes. Through the process of network governance, local nodes and global circuits have fed each other. For this reason, local places of worship such as churches, synagogues, or mosques can no longer be studied comprehensively outside their attachment to the global circuits that influence and shape their local behaviors.

Although network governance is nothing new in the administration and global management of these scriptural religions, each has adopted various technologies of communication to interact with its distributed web of nodes. More recently, these faith communities have domesticated the Internet to serve their religious, administrative, and humanitarian purposes.

Since digital technologies have become part of their infrastructure of communicational operations, the book analyzes their role in the mechanics of the global leadership and administrative structure of these religions. IT (Information Technology) has permeated various aspects of daily life, personal and institutional, including the management of global governance. In its performances, network governance has become digital. As such, it is also both transglobal because of the worldwide distribution of its nodes and mobile because the entire circuit contributes at one time or another to the governance of the global network.

Digital network governance is conceived of as a coproduction resulting from the interface or interaction between digital and non-digital modes of communication choreographed on a global scale. The issue of governance provides a way to examine how nodes affect and are affected by global networks. The shift in the study of religion from a focus on locality to an emphasis on globality assumes that much can be learned from studying the modes of network governance of these multinational organizations. The embeddedness of the Internet in network governance adds value, mobility, and efficiency to the process of governing. This book on network governance and its digital embeddedness shows how the latter enhances the performance of the former.

Why Jerusalem, Rome, or Mecca? The centrality of Jerusalem as a key site in the global governance of the Jewish faith is based on its spiritual and theological meaning as a holy city, the materiality of the religious symbols that it houses (the First and Second Temple, King David's Tomb, and other historical remains), and the global movement of pilgrims that it generates yearly. The Temple (now the Wailing Wall or the Western Wall) has been the main point of attraction for residents, pilgrims, and tourists alike because of the rabbinical injunction before the dispersion for the faithful to pray there one to three times a year.[1] Jerusalem also gains its centrality in Judaism and Jewish history because of the early prescriptive obligation imposed on Jewish families to pay a religious tax for its upkeep and because it serves as the seat of governance of the rabbinical leadership. Moreover, Jerusalem has been a symbol of the longing of the diaspora for the ancestral homeland, as exemplified by the term of address "Next year in Jerusalem," which was popular among Central and Eastern European Jews before World War II and the establishment of the State of Israel.

In addition to its centrality to Judaism, Jerusalem is an important site for Christianity and Islam. For Christianity, Jerusalem is the site of the death and resurrection of Jesus, and countless Christian pilgrims flock to the city every year. For Islam, Jerusalem is the site of the ascent to heaven of the Prophet Muhammad and of the third holy mosque (the al-Aqsa Mosque and the Dome of the Rock), which brings Muslim pilgrims to the city.[2] Jerusalem thus is a holy city shared by the three monotheistic faiths because it contains key parts of their religious heritage.[3]

For Christians, the physical site of Jerusalem (including the Church of the Holy Sepulcher, the Cenaculum or Last Supper Room, Dormition Abbey, Saint James's Cathedral, and the Gethsemane Basilica) is not the only symbol that holds importance in the expression of their faith. However, Jerusalem combines with the mystical side of Christianity to complete the global body of the Church. Early on, the Pauline doctrine distinguished the "earthly Jerusalem" or "Jerusalem of the flesh" from the "heavenly Jerusalem" or "Jerusalem of the spirit."[4] This distinction recalibrates the topographical geography of the faith by addressing its universality and by unveiling the globally inclusive nature of Christianity's message.

As a consequence of its genuine importance to the three faiths, Jerusalem is a contested holy city. This competition figures in the mathematics of the network governance that each religion upholds. It is a struggle for domination, control, and access to ensure that their religious sites are protected and available to pilgrims.[5] The dynamics of the contest, whether Jews, Christians, or Muslims are in charge of the political administration of the city, has varied over the years.[6] It also varies periodically, not only because the pilgrimage economic cycle enhances the income levels of the residents of the city, but also because it brings pilgrims who expand the global fame of Jerusalem. Similar dynamics happen in the cases of Rome and Mecca.

Mecca is the holiest site of Islam and constitutes a central node in the global circuit of Muslim places of settlement and worship.[7] It gains its sacredness through its special role in Islam, because it is believed to be the location of the Garden of Eden, because the Ka'ba, the first place of worship built by Abraham,[8] is located there, and because it is the birthplace of the Prophet Muhammad. Its centrality is reinforced by the injunctions in the Koran that command the faithful to face Mecca when in prayer and to journey as a pilgrim to Mecca at least once in one's lifetime. In addition, it serves as a diffusion center for the spread of the orthodox interpretation of Islam, the dispensation of doctrinal guidance, and the regulation of personal conduct.[9] No other site has combined so many religious characteristics that symbolize the unity,[10] diversity, and globality of the faith.

Rome is a different story. Unlike Jerusalem or Mecca, it is neither the place of birth of the central figure of religious devotion, the site of the primitive Christian Church, nor the place where the most important patrimonial heritage of the religion is located. Its primacy comes mainly from the martyrdom of Peter and his burial on the Vatican Hill in Rome, because he was selected by Jesus to be in charge of the nascent Church. His mission was clearly spelled out by Jesus in his "Tu es Petrus" dictum, which defined the *primus inter pares* position of Peter among the apostles.[11] The Petrine doctrine that was later developed by the Church establishes the primacy of Peter over the other apostles and the principle of "apostolic succession."[12] This provides the rationale for the centrality of the Vatican (i.e., the Holy See) in the global governance of Catholic Christianity.

Jerusalem, Rome, and Mecca thus are identified as central sites because they are representative symbols of Judaism, Catholic Christianity, and Islam. They are recognizable nodes with their own quality, religious topography, and theological meaning. Since the focus of this study is on network governance, various other nodes that constitute its landscape also are invoked as examples to show their contribution to the performance, behavior, and integration of the total circuit.

It is important to address briefly the issue of data collection or methodological strategy. I started with the assumption that one may penetrate a global circuit through any of its local nodes, and that studying it from the standpoint or positioning of any node has its advantages and disadvantages. The data were collected from research carried out in Rome, Jerusalem, Paris, Berlin, London, Montreal, San Francisco, and Port-au-Prince. The goal was to observe and interview or make requests for clarification from religious leaders (priests and bishops, in the case of Catholicism; rabbis and administrators of Jewish institutions, in the case of Judaism; and imams and individual Muslims knowledgeable in Islamic jurisprudence, in the case of Wahhabi Islam). Of course, a good portion of research time was allocated to familiarizing myself with the available written secondary literature for the purpose of reinterpretation.

Various websites of different nodes of these religions provided raw data for further interpretation. I consulted the websites of the major headquarters of various branches of Judaism, including Orthodox, Conservative, Reform, and the Lubavitch Movement; Catholicism, the Vatican; and Islam, including the Ministry of Hajj in Riyadh, the Islamic University of Medina, and civic organizations such as WAMY (the World Assembly of Muslim Youth, a division of the Muslim World League) in Jeddah, the Religious Affairs section of the Saudi Embassy in Washington, D.C., and websites of organizations sponsored or financed by the Saudi government in North America and Europe. I interpreted these data in light of religious prescriptions from the Torah, Bible, or Koran and legal or moral injunctions put forward in halakhic, canon, or shariah law.

There is added value in comparing the governance apparatus, structure, and operation of these Abrahamic and monotheistic global religions, as each exhibits features singular to its extraterritorial deployment and everyday practices. This is a limited undertaking, however, because the analysis focuses on network governance and not on other factors such as origins, doctrines, traditions, or interactions with state and civil government. The overseas expansion of these religions has led each to develop faith communities across the globe, establish mechanisms of communication so as to facilitate exchanges between congregations, forge new administrative structures in each territory, and engineer new forms of global collaboration, solidarity, and mutual aid. The everyday operation of the widespread connections between these entities, which vary in social importance, yet pursue similar goals in diverse locations, requires a network management and administrative structure of global governance that reflects the contribution of each site to the social reproduction and structural viability of each global religious regime.

Though it may seem problematic to compare a centralized religious institution such as Catholicism with two decentralized ones such as Judaism and Islam, it is possible to examine the network governance of these three global religions despite differences in their internal make-up, structural organization, and geographical deployment. Catholicism exhibits a more centralized, hierarchical, and regulatory form of governance, whereas the other two religions are more decentralized, neither having a joint headquarters that provides religious guidance, polices the boundaries of the doctrine, ensures compliance by the membership, and oversees the expansion of the faith.

This comparison is also undertaken despite recognizing different authority structures and systems of regulation endorsed by different groups within a given tradition. Not only are authority structures different between religious communities of different faiths, but they are also different within religious communities of the same faith. For example, in Islam, Sunni, Shiite, and

Sufi Muslims adhere to different religious traditions, adopt religious practices sanctioned by the religious leaders of their branches, and expand their reach beyond the confines of the national borders of their states. Each group in Judaism follows the teaching and guidance of a different sort of "rabbi," who provides leadership to the religious movement and shapes its own "kosher tradition." In Catholicism, the variability of authority structures and local bureaucracies is based on cultural differences and different religious traditions within the same faith, such as local bureaucracies established by the Coptic Catholic Church in Egypt, the Armenian Catholic Church in Lebanon, the Chaldean Catholic Church in Iraq, the Syro-Malankara Catholic Church in India, the Greek Byzantine Catholic Church in Greece, the Syriac Catholic Church in Syria, the Melkite Greek Catholic Church in Syria, and the Maronite Church in Lebanon. These are all Catholic churches in communion with Rome, but with different canonical regulations. Thus, heterogeneity of nodes is a characteristic of network governance.

The lack of symmetry among these three global religions, however, is not an impediment to comparative analysis. The presence of a central node does not imply the absence of a network arrangement; rather it calls attention to the singularity of this form of network governance. The argument developed here is that dissimilar things have also been the objects of comparative sociological analysis for several decades. One engages in comparative studies not simply to unveil similarities, but also to uncover differences. In the social sciences, one generally compares not only similar realities, but also different structures. In the first scenario, the goal is to explain how similar structures can exhibit differences in their modes of operation and to highlight how these findings or insights can be useful in correcting inadequacies and enhancing performances. In the second scenario, the goal is to explain how different structures could exhibit similarities in their modes of operation and in the process deepen our understanding of the network governance of global religious regimes.

This comparative analysis highlights not only the chasm between Catholicism on the one hand, and Judaism and Islam on the other, but also pinpoints differences between the two decentralized religions. In this instance, decentralization does not mean structural similarity, but rather explains the idea of a plurality of differences and the recognition of different modes of organization. In addition, centralization does not mean total absence of decentralization. In other words, instances of decentralized practices may emerge in a centralized structure of governance in the same way that examples of centralized practices may pervade in a decentralized structure of governance. A comparative approach allows us to ascertain whether these observations can be attributed to incidental occurrences or are intrinsic to the performance of formal systems. A similar argument has been addressed earlier in the case of informal systems.[13]

In this book, a comparative analysis of network governance of these three religious regimes is undertaken to shed light on five topics of importance.

These include the regulation, control, and supervision of cross-border mobility of clergy, religious teachers, and seminarians; direct or indirect oversight in cross-border circulation of church money; global management of church infrastructure (shrines, church buildings, places of devotion and worship in the holy lands); global protection and security of church property and practitioners of the faith; and general guidance over issues of regulation, commercialization and consumption of dietary products (kosher foods for Jews, halal foods for Muslims). Diverse nodes are involved in the network governance of these activities and each religion has developed a network administrative structure to manage or oversee these operations or chunks of the unfolding process across the globe.

Despite their diverse modes of operation, Catholicism, Judaism, and Islam all exhibit forms of network governance that are engaged in cross-border practices with transnational actors in distinct geographical nodes. Although these networks display unique social architectural contours, they share similar features in terms of engaging religious leaders in different countries, securing the wellbeing of the network, and fully projecting their global identity rather than local expressions of the faith. Comparing network governance is a key mechanism utilized in this analysis to unveil the logic that promotes, permeates, and sustains their differences.

In a systematic attempt to explain how global religions proceed in governing and administering their global networks of sites and religious communities, three empirical chapters analyze and document the specificity of the network governance process undertaken by each of the three monotheistic religions.

Chapter 1 reviews the vast areas in which these religions have developed and implemented procedures to ensure that dispersed congregations remain in communion with the larger network, collaborate with each other, and help those in need wherever they might be located. To explain the dynamics of this form of governance and to unveil the unique characteristics of its deployment by each faith, the chapter begins by rethinking key concepts such as "global networks," "global religions," and "network governance" in order to explain their genealogies and precise meanings in the sociological and political science literature. In other words, it presents a theoretical frame of reference that helps the reader to understand the global expansion and rationale behind the mode of network governance that each religion has developed. The chapter also briefly describes the three models of network governance espoused by Judaism, Catholicism, and Islam. It further discusses the mobility factor, enhanced by the use of digital technologies that make the circuit operative in real time.

From an analytical standpoint, chapter 1 explains a shift in emphasis from a methodology that focuses on nation-states to a methodology that acknowledges the importance of transnational and global networks. The global network, and not one particular node, is the object of study or the focus of analysis. The structure of the book reflects this preoccupation with the global level. The emphasis is on comparing transnational blocs, instead of three prominent nodes. In other words, how network governance operates in these three religions is the pivotal question to which the book attempts to provide an elaborate, disciplined, and thoughtful set of parameters and theoretical explanations.

The study of global Judaism and its network governance is the focus of chapter 2. Unlike the Catholic Church, Judaism has no central authority that decides and orients the spiritual and moral direction of its global community of believers. Instead, it has several heads, such as the chief rabbinate in the case of Israel, England, or France, and the leaders of branches of the faith, including prominent rabbis of each sect of the Haredi (ultra-Orthodox) community such as the Satmar, Ger, Belz, Bobov, Puppa, Vizhnitz, Munkacs, and others.

The chapter develops a frame of analysis to explain what network governance means in the specific context of decentralized governance and how it evolves and operates. In other words, it asks, "What is its procedure of practice?" Two types of networks are of interest here: the global circuit that constitutes the *religious identity* of a branch of Judaism and the larger network that at times connects both practitioners and secular Jews in the expression of their sense of *peoplehood* or *ethnic identity*. Accordingly, these two lines of inquiry analyze the multiple globalities that traverse Judaism and how the local synagogue must be seen as a global node. In this light, it was necessary to use the synagogue, religious associations, and the state, as mechanisms in the production of a new typology. This contrasts with the older typology of Conservative, Orthodox, and Reform Judaism that stresses differences in ritual structures and contents, based on doctrinal orientation, or Sephardic or Ashkenazi background. In these terms, the chapter attempts to spell out the variable architecture of global Judaism.

The chapter identifies five institutional practices for empirical description and analysis according to how they feed the mechanisms of global governance. First, the global commercialization and distribution of kosher products, which accounts for their global mobility and consumption; second, the global mobility of religious leaders, including rabbis, religious teachers, and rabbinical students, which sheds light on the infrastructure that supports the expansion and strength of Judaism; third, a global security mechanism which helps protect the synagogues and congregants and facilitates exchange among various diasporic enclaves for greater surveillance capability and efficiency; fourth, the policing of the borders of the faith which brings various branches of Judaism together to confront the propagators of other faiths in order to protect and preserve the religious patrimony and heritage; and fifth, the significant role the State of Israel plays in the global governance of the faith.

Chapter 3 investigates the network dimension of corporate Catholicism. It shows that, whereas the emphasis in previous studies has been to highlight the corporate structure of governance in Catholicism, with the Vatican bureaucracy as its representative symbol, the institution has always operated on the basis of network arrangements and relations. The difference, when contrasted to less integrated multinational organizations, is that the presence of the Vatican as a central locale both influences and is affected by the network and that it occupies a central position in the global management of the faith. The chapter explains how over the years, the Church has developed a global infrastructure (including parishes, bishops' offices, lay associations, religious orders' mission-churches, and sister churches) that allows its smooth functioning and social reproduction on a global scale. It analyzes the spatial materiality of the Church's network governance in terms of such units, their positions in the global network, and their relations with the Vatican.

Some aspects of this global network structure receive closer scrutiny. For example, the global mobility of the clergy makes it possible to assign people to posts where their talent is most needed, so that communities that are unable to produce their own clergy will be provided necessary religious leaders. The training of the younger cadre for global leadership takes place in Rome by enrolling them in Catholic universities or schools of theology so that they can meet the needs of their own religious communities and those of the global network. The circulation of monetary contributions and their distribution to communities in need to pay for overhead and Church infrastructure and to use for the upkeep and maintenance of the patrimony, especially in the holy lands, necessitates a global management of assets and revenues. The security mechanism that can be deployed to protect the pope at home and abroad is also an aspect of network governance that has global implications and dimensions. At the same time, the protection and preservation of shrines in the holy lands and the rest of the world requires the attention of several nodes in the global network in terms of political influence, financial contributions, governance, and administration. Outsourcing, whereby a central node delegates another node to be in charge of one of its functions, shows the flexibility and collaboration among nodes that make up the global network. As is the case in all three religions, digitization has expanded these global contours of network governance into the virtual realm.

The analysis of the distributed cellular network governance of global Islam, with a particular emphasis on Wahhabi Islam, is the focus of chapter 4. After a brief introduction explaining how Saudi Arabia became the headquarters of this confessional branch of Islam and how Mecca is central to the global practice of the faith, it discusses the distinct nature and characteristics of Wahhabi Islam. In so doing, it pays attention to the interpretations of both practitioners and critics. It further elaborates on the distributed cellular network governance frame of reference to explain its content and its similarities and differences vis-à-vis other forms of network governance.

To explain how distributed cellular network governance is deployed, the chapter discusses the role of the state in the geographical expansion of the faith and the global strategy of proselytism that it develops and uses in order to achieve its projected goal. It describes and examines various policy and procedural practices established conjointly by religious leaders and the government of Saudi Arabia to discipline and control the global mobility of Muslim bodies. This technology of control is used in reference to the performance of the hajj and minor pilgrimages, the training of overseas students at the Islamic University of Medina, and the assignment of religious educators and leaders to missions abroad.

Furthermore, in an effort to understand the interactions between nodes in the global circuit, the chapter explains the use of the mass media by religious leaders to propagate the faith outside the geographical borders of Saudi Arabia. This includes short-wave radio programs, satellite television, and, more recently, the Internet. The latter provides direct interaction, in real time, between the Ministry of Hajj in Riyadh and the ministries of religious affairs in other Muslim countries for the purpose of cross-border surveillance and national security. What comes out of the process of analyzing the global linkages between Saudi Arabia and the worldwide nodes of Wahhabi Islam is a new vision of the way in which network governance of this global religion operates and of the constant upgrading of the infrastructure of the kingdom to meet the needs of the mobile bodies (pilgrims) that inhabit the global circuits of nodes.

The concluding chapter addresses the issue of the differentiation of the global public, which each religion defines in a unique way—that is, the Jewish *kehillah*, the Catholic Christian *ecclesia*, and the Muslim *ummah*. The main focus of the chapter is on delineating key variables that constitute the core aspects of network management and governance. The chapter spells out the identifiable features of network management in terms of the multisite organizations of global religions, the loyalty of sites to the network, the outsourcing of functions, the role of the network as an incubator of innovation, the transnational flows of knowledge, the changing status of the nodes, and the whole and complex issue of multiple administrative cultures and network management diversity. All of these elements are invoked to show the centrality of network governance and management in the deployment and social reproduction of these global religions.

1 Network Governance of Global Religions

For many years, Jerusalem, Rome, and Mecca have played important roles in the lives of the faithful throughout the global landscape, both directly and indirectly.[1] As global cities, they contain religious and sacred sites that attract pilgrims every year, serve as official or symbolic headquarters of a global religion, are used as a major or the principal seat of governance of a multinational religious organization, and function as a central spiritual or administrative node of the global network of Jewish, Catholic, or Islamic practitioners. Although the history, geography, institutional organization, and theological meaning of these sites are well investigated and documented, very few attempts have been made thus far to compare their forms of governance and inscriptions in the global networks of nations and states apart from their role as holy cities.[2]

Through missionary work and as a consequence of immigration and diasporization, each of these monotheistic religions has expanded beyond its national borders as a way to spread and share the faith with potential overseas converts.[3] The Catholic Church, in particular, has been the most successful in its global outreach and has developed, in the course of its history, an entire department in the Curia in Vatican City devoted to missionary endeavors. This age-old practice of proselytizing in faraway lands led to the creation of a global governance structure that allows the central administrative site to develop a hierarchical organization that networks with, interacts with, and provides leadership to satellite sites.[4] Just as Jerusalem represents the principal site in the practice of Judaism at home and abroad because of its geographic centrality to the faith, Mecca represents the center of Islamic faith. Muslims face toward Mecca during daily prayer performances and are required to participate in a pilgrimage to the holy city at least once in their lifetime. In this sense, each of these holy cities contributes in many different ways to the governance of a global network and, therefore, constitutes a privileged site for the study of the globalization process, in general, and network governance, in particular.

The exegesis of the global inscription of Jerusalem, Rome, and Mecca will be carried out in this book from the perspective of "network governance" in an effort to compare how each undertakes this essential task in daily practice. Under this rubric, I will stress and investigate three issues: the administrative governance that sustains, reproduces, and expands the interpretation and practice of each faith; the spread of each faith and religious movement throughout the world; and the global infrastructure developed for the network governance of this global religious formation. In the process of transglobalization, these three dimensions are intertwined in such a way that they tend to sustain each other. For example, a historical religious event ties a religion to a physical place and transforms it into a sacred site that commands the attention of brethren located in diverse national sites. The specific place is significant because it becomes the site of an epiphany that feeds subaltern and satellite sites spiritually and because it creates a meaningful physical anchor as a point of geographical reference for the religion or religious movement.

It is my view that for too long, the comparative analysis of the three monotheistic religious systems—Judaism, Christianity, and Islam—has been conducted from a doctrinal standpoint, in terms of the similarities and differences of their theological contents and hierarchical clerical architectures. This top-down theological approach is limited by its normative doctrinal orientation. In contrast, this study proceeds from a different angle and analyzes the institutional infrastructures of these multinational organizations that make networked global governance possible, manageable, and efficient. The organizational infrastructure is a key factor that may enhance or impede the practice of good governance, tolerance, and ecumenism.

A series of mechanisms, not all of which are common to Judaism, Catholicism, and Islam, affect global governance at a central site. These administrative seats of governance serve as the depository or organization endowed with authority to preserve the covenant. This is accomplished in myriad ways: through the policing of orthodoxy and orthopraxis of adherents worldwide by silencing theologians, criticizing or even banning unorthodox practices, and upholding ritual practices; through expansion of the faith by proselytizing with the help of missionary work in foreign countries; through the multinational make-up of institutions such as the Curia and the College of Cardinals who elect the popes; through the *encadrement* of immigrants by transferring homeland priests, imams, or rabbis to take care of them while living in diaspora; through Roman universities (the Gregorian University in Rome, for Catholics), Islamic universities (the Islamic University of Medina, for Muslims), or rabbinical seminaries (the Schechter Institute of Jewish Studies in Jerusalem, for Jews) for the training of future leaders of the religious institution; through the transnational transfers of cadres (assigning a foreign bishop to the Curia, a Curia bishop to a foreign diocese, an imam or a rabbi to an overseas congregation, moving nuncios from one territory to another); through the pope's "Urbi et Orbi" blessing, on Christmas and Easter Day, to the city of Rome (in his capacity as the bishop of Rome) and to the rest of the globe (in his capacity as the pontiff of the universal church); through the collection, management, redistribution, or reallocation of finances; through the protection of religious sites in the holy cities or lands, the maintenance of old sites, including pilgrimage centers, and the building of new churches, mosques, or synagogues; through the definition of the *doxa* by councils and synods; through guidance on moral and political issues on world affairs addressed to Christians, national governments, and international organizations; through the periodic publication of papal encyclicals; through a centralized corporate (Catholicism), network (Judaism), or distributed cellular (Islam) governance structure; through the establishment, staffing, or financing of religious organizational structures in foreign lands; through disciplinary actions (the dismissal of priests and bishops, or the excommunication of foreign dignitaries)[5]; through the system of religious taxes (tithes) imposed on or voluntarily given by the brethren for the maintenance of infrastructure, geographical expansion of the faith, defense against individual groups or state detractors or persecutors, sustaining the work of the charitable organizations of the faith, and financial redistribution to communities in need; through the religious legal system established to sustain the faith and resolve conflicts or disputes (canon law for Catholics, shariah law for Muslims, or halakha law for Jews); through the lunar-solar (Judaism), solar (Catholicism), and lunar (Islam) calendar systems (the Gregorian calendar, Muslim calendar, or Jewish calendar) that differentiate these religions' tempo from that of the civil week and that provide a cadence to religious life; through the land they own throughout the globe; through the hierarchy of positions they establish and that is reproduced in every country in which they operate; through the newspapers, radio and television programs, and websites they operate or sponsor; and through the liturgical or ritual practices they endorse or impose on local synagogues, churches, or mosques.[6]

Governance of the global network implies different modes of practice, depending on the hegemonic or subaltern status of the religion in society, the time period, and historical or contextual circumstances. Such differences include, for example, different methods undertaken for the expansion of the membership of the faith, whether through recruitment (proselytism or forced conversion), competition, annexation, subjugation, or elimination of other religious bodies (through colonization, crusade, or concordat); different strategies for the protection of some sites, whether by paid or volunteer security forces and collaboration with the municipal police or other means, different methods for achieving the global coordination of representative agents and systems (apostolic nuncios, the appointment of bishops, rabbis, or imams, annual visitations via delegates); and different regional or hemispheric bodies or organizations that mediate between the global network and local units or between the administrative religious center and its peripheral offshoots in different ways.

This study documents and explains how the organization of infrastructures is a key factor in understanding the global governance structure of these religious institutions; how these religious bodies operate as multinational corporations with a religious and missionary purpose that allows them to help, educate, and evangelize; how they diversify their techniques and procedures while maintaining the same doctrine throughout the globe; and how they have shifted from transnational and multinational postures to global and multicultural postures. For example, the Catholic Church at first used the same rituals, language, and liturgical musical instruments everywhere; since the Vatican II, it uses indigenous languages and musical instruments and accommodates different national traditions, even in matters related to fasting days. In other words, global governance in the history of these religious institutions has evolved from a homogeneous to a heterogeneous structure of practice. In all of these scriptural religions, network governance has contributed both to the splintering of these religious institutions and to collaboration between diverse groups.

NETWORK

A network is an interconnected set of units constituting a distinct aggregate or forming a pattern sustained by the mode of relations that incorporate them in the same universe. Networks, which depend on the geographical locations of units, the goal a local group wants to accomplish, or the use of an infrastructure, can be local, regional, national, transnational, or global.

In the scholarly literature, network analysis has been put to use in a variety of ways and has been the basis for the development and application of a number of theoretical models. One can think, for example, of random graph models, small-world models, scale-free network models, and preferential attachment models as developed in such diverse fields as biology, physics, mathematics, chemistry, and sociology.[7] Furthermore, network has been invoked as a metaphor, a method, a theory, and even a science, "the science of networks."[8]

The basic features of network analysis that physical, biological, and social scientists have grappled with are many.[9] Analysts have used the network approach to attempt to resolve issues of connectedness between nodes, establish group parameters, unveil the trajectories of flows, interpret the intensity of relational ties, identify nodes, evaluate actor centrality, map communication patterns between nodes, assess the formation of clusters, and analyze relations between clusters. The network approach stresses the importance of focusing on relationships between nodes as a way of unpacking different aspects of social reality.

The network perspective provides an ideal method to uncover the matrix of relations between nodes; define the boundaries of clusters of nodes; follow the direction of communicational, monetary, and travel flows between nodes; and unravel the making, unmaking and remaking of nodes.

GLOBAL NETWORKS

Global religions operate on the basis of the global networks that constitute their membership and that stretch their activities beyond the borders of the territory where they first appear. Linkages between dispersed nodes form these global networks and inform us about the geography covered and the types of activities in which the rank and file are engaged.[10] The globality of such networks must be seen in terms of the universalism of the faith as it is welcomed and accepted by people across the globe, in terms of its geographical spread, but also in terms of the relations these various nodes maintain among themselves.[11]

Global religious networks have their own peculiarities because each follows specific ground rules, partakes in the same religious ideology, and remains prone to expansion, diversification, a multiplicity of relations, and reproduction over time. These global networks are polyvalent in the sense that they comprise more than one single network and have many strands. One also may distinguish inward from outward networks and dormant from active networks.[12]

Inward networks deal with activities within the membership of the faith and can be local, regional, national, transnational, or global. These networks are established for various reasons, such as to collaborate on projects, to enhance learning, or simply to help brethren in need. Internal networks involve the vibrancy of relations among members of the faith, irrespective of their place of residence.

In contrast, outward networks address issues of the relations of nodes with groups of people who do not belong to the faith or do not hold the same interpretation of the faith. Most established religions maintain such networks to sustain a dialogue that can benefit all parties involved, to collaborate on issues of importance to multiple faiths, or to participate in humanitarian endeavors.

Both inward and outward networks can remain permanently active, be cyclically or sporadically in operation, or be dormant at times.[13] This is a way of saying that global networks have their own life cycles, and any analysis of their behaviors must attend not simply to space or geography, but also to time and history. These vary from one global religion to another in the deployment of each's tentacles, the composition or multiplicity of nodes within a given network, and the contents of the relations between sites.

Global religious networks function on formal, informal, and virtual bases and can operate using one or all of these means of communication. The myriad ways in which networks intersect each other or form specialized aggregates reflect the complexity of multiple webs of relationships and the existence of plural global circuits. Clearly, all nodes in this universe do not have the same weight and therefore do not influence the circuit in the same fashion.

Global networks are useful for the global sustenance of a faith, not only for communication among members and the sharing of resources, but also because they feed or create conflict that must be managed globally. The legal system that each global religion establishes provides a frame of reference for the resolution of conflicts generated within its internal or external networks. Therefore, the global governance of the networks includes the governance of the borders of the networks, including tensions and crises that may develop as a result of the relations of the network's nodes with external entities.

MULTIFORM NODES

Nodes display diverse characteristics, depending on their position and weight in a global network. A node may be part of more than one subnetwork, and its role depends on whether it is called upon to provide "narrow" or "broad" performances. Global religious networks operate with controlled, autonomous, and interdependent nodes. The Catholic Church is a good example of a global network with controlled nodes. Its headquarters and worldwide structure are hierarchically organized, and its global networks function within these constraints. In contrast, Judaism provides a model of embedded nodes. Here, each religious movement is autonomous but may collaborate with others to discuss and fix problems internal to the cosmonation, to propose solutions to external problems, and to enhance the security of the State of Israel. Islam presents yet another model, that of interdependent nodes within global circuits. These transnational nodes compete against each other to attract more converts and resources to their groups, but they also collaborate with each other, for example, for the purpose of sharing the sacred sites of Medina and Mecca in Saudi Arabia.

The global network concept must be understood from the standpoint of multiple and multiform nodes that constitute the landscape of practitioners. By virtue of the linkages they develop and sustain among the faithful, nodes behave differently and fulfill different tasks. In light of this observation, it is clear that a global network of nodes does not imply uniformity of practice, but rather a multiplicity of relational arrangements. Therefore, it is necessary to study the mode of operation of some nodes to understand their relational linkages and differential behaviors.

Saudi Arabia, an important node in the global networks of Wahhabi Islam, provides us with some clues regarding the multiple ways in which nodes may operate. The Saudi religious authorities intervene in the affairs of overseas nodes either directly, through their nongovernmental organizations (NGOs), or indirectly, through the Saudi government.[14] Their activities include building mosques, supporting Islamic centers, and financing research institutions and scholarly activities in Muslim and non-Muslim countries. In Pakistan, Saudi Wahhabis fund *madrassa* schools, whereas in Turkey, they finance dormitories for students who cannot afford rent. These are some examples of how different Wahhabi Islamic strategies to enhance proselytizing networks can result in multiform nodes.

In Judaism and Islam, no single node controls the global network, but each node has the potential to influence its direction in one way or another, because they are participants of equal status. In theory, the actors who constitute the nodes of these global religious networks are "relative equals,"[15] and their relation to one another must be characterized by "mutual resource dependency."[16] In short, they are dependent on and responsible for the well-being of one another. However, power and wealth may make some nodes more influential than others.

Collaboration is the main means used to help any node in the network and the main strategy for nodes to overcome their local problems. The very existence of a global network depends on successful collaboration among nodes to make its operation sustainable.[17] Of course, some nodes are more likely than others to be partners in collaboration as a result of need, location, cultural practices, and linguistic affinities.

MULTIPLE RELATIONS

Relations between nodes are neither one-dimensional nor always smooth, but rather are pluridimensional and pregnant with tensions. They influence each other and engage in packaging or repackaging their strategies in order to achieve their goals in the network. The interpretation of a theological doctrine may evolve not simply as the result of the passing of time and access to new knowledge, but also due to the repackaging of the interpretive skills or traditions of local nodes.

Relations between nodes may remain dormant until they are activated for the purpose of resolving a specific issue. Active interaction between nodes may be temporary, cyclical, or permanent, depending on local needs. For example, sometimes the network operates on the basis of temporary collaborative dyadic relations between nodes. For these reasons, network governance is contextual. In this light, the concept of the "ecology of governance"[18] is useful because it helps us identify the contextual setting in which each religion operates and manages its global network. In this specific context, one may think of the administrative parameters (what is to be governed), the human-resource parameters (the leadership structure of the religion), the scriptural parameters (the limitations set by doctrine), the geographical parameters (the areas covered), and the boundary parameters (the relations with other entities, including other religions). This ecological infrastructure points to the diversity and multilevel forms of the network governance of global religions.

GLOBAL RELIGION

The concept of a global religion has been defined and given meaning in a number of different ways.[19] In scholarship and debates on theology, spirituality, and the history of religions, the concept of a global religion has been used to differentiate such a religion from minor faiths or to claim its central importance in world affairs. Global religions are seen as those in which people all over the world recognize the truthfulness of the faith, in contrast with local religions, which have no influence beyond a certain region or territory.[20] A religion can also be seen as global, not only because of its geographical spread, but also because of its inclusiveness, as it encompasses everything in the universe (animals, plants, planets). This perspective highlights the spiritual dimension of the faith.[21] The qualifier "world" is also added to religion by historians of religions in order to differentiate minor religions from those that have a large impact on the international community.[22] The thrust of this approach is to present a typology of religions according to their importance in world affairs.

Most of these accounts of global religion emphasize the universality of the faith, its geographical expansion, and how globalization has affected religion.[23] In this analysis of network governance, a religion is said to be global when it establishes itself in areas beyond its national boundaries and maintains ongoing connections with these satellites. There are several means by which these satellites are developed: conquest and colonization, immigration and diasporization, and missionary projects. Once satellites have been developed, by whatever process or processes, religious leaders may be trained or appointed to lead the diasporic group and its new converts. Judaism is a covenant between God and the people of Israel, but the religion became global not because of the conversion of Gentiles, but because of persecution that led to emigration and diasporization. Therefore, it is global because of the effects of persecution.

A religion may establish itself to be a universal phenomenon at its inception, as in the cases of Christianity and Islam. Although developed locally, these religions aimed to be global. The message they offer is not to one country or people, but to the globe. It is not, per se, the geographical spread that makes these religions global, however. It is their global orientation that has led to extraterritorial expansion. Christianity and Islam are global by orientation.

A religion is global when it defines itself as a multinational, multicultural, and multiethnic faith-based community project, as reflected in its demographic composition. The *ummah* is not solely the community of Muslim believers, but also of the "People of the Book." The plural composition of the group accounts for the international dimension of the faith.

The process by which a religion becomes global or globalizes itself varies extensively.[24] In the case of Catholicism, globalization has occurred as a result of missionary endeavors and the imperial expansion of European states in the sixteenth century; in the case of Judaism, it occurred as a result of persecution and diasporization; in the case of Islam, globalization has taken place through the expansion of trade routes, conquest, and the imposition of a religious regime and lifestyle upon the people of newly occupied territories.

The globalization of an existing religion occurs when a faith spreads worldwide and in the course of its transnational itineraries readjusts its teaching, rituals, and practices as a consequence of its multiple relocations. A global religion also results from the effect of the transformation of the external world onto its internal structure. The religion becomes global not simply because of its own internal dynamics, but because of the effect of external forces on its organizations. These externalities contribute to its reshaping. These two processes are emblematic of the inward and outward forms of the "globalization of religion."

A faith also can be born out of, or as a creation of, the globalization process. Such a faith emanates from a different process than that of the globalization of religion. Global society either gives birth to such an entity or provides the context and milieu for it to occur.[25] These are multicultural religions that attend to the needs of multicultural communities. In other words, it would be unthinkable to imagine the existence of such a faith outside a global context. The term "religions of globalization," that is, religions born out of the transformations brought about by globalization, indicates the origin and itineraries of the global contents of such religious systems.[26]

A religion also can be seen as a "carrier of modernization and globalization."[27] The religion fuels various aspects of society in a way that makes globalization attainable. It provides operational legitimacy and is the site that generates the early elements and the rationale upon which the international modern legal system is built. It energizes the world in a way that necessarily leads to the path of globalization it now enjoys. This vision of global religion is referred to as "religion in globalization."[28]

The global is constituted differently in the case of each religion because each covers specific sites in which its practitioners live and over which each has jurisdiction. In the case of Judaism, the global element of the faith refers mostly to people of the same ethnic background, both at home in Israel and abroad in the diaspora, that is, the global Jewish cosmonation. For Christianity and Islam, the global element includes people with many different ethnicities and nationalities who belong to different nation-states. This observation simply restates the view that multiple global religious networks partition the global landscape and distinguish one religion from the other.

Global religion from the standpoint of governance is the outcome of several processes: an interpretation of Scripture to allow others who reside beyond the state's boundaries to join the faith, dispersion to reach people in other communities (migration, conquest, mission), and the implantation of a local structure of governance to organize the new brethren and to ensure a line of communication with the headquarters or the sponsoring church. In other words, global religion requires the creation and maintenance of a global structure of governance. A central issue that permeates the theoretical orientation of this book is how a structure of global governance choreographs the global network activities of the faithful.

NETWORK GOVERNANCE

As previously noted, the global governance of each religion has its own singularity and characteristics that reflect the ways in which it deploys activities and the geographical dispersion of its membership. Judaism, Catholicism, and Islam are each involved in some form of network governance, with or without a common headquarters, just as each becomes involved in some form of hierarchical organization. In fact, the very concept of network implies a hierarchy of status in its nodes, because they are unequal partners (in terms of wealth and power) and therefore not identical. That inequality is seen in the hierarchical model of the Catholic Church, but also in the models of independent or autonomous units that constitute both Judaism and Islam, because some nodes have more weight than others in the voluntary network in which they may occasionally participate in to resolve issues of common interest to the larger community.

The unique forms of global governance found in Judaism, Catholicism, and Islam use different types of networking practices in their operational structures.[29] The members of each faith, first of all, commonly contribute to the ongoing process of designing the network to reflect the values of the religion. These religious webs are not networks developed for profit, but rather they are developed to proselytize and sustain the faith in its multiple locales. The second common feature they share is network evolution, which means that all nodes have developed characteristics peculiar to certain historical periods. For example, some nodes have developed as a result of an earlier period of colonization, persecution, or voluntary immigration. Third, members of each network share the same faith, respond to a religious call, and willingly contribute to and participate in it for their personal spiritual growth and the well-being of the religious group. The fourth feature they have in common, as self-sustaining global networks, is that each node successfully contributes good deeds, prayers, professional skills, and reverence for the faith.

For many years now, social scientists have used network analysis as a methodological tool to understand the behavior of social systems. In this light, one may identify five generations of network research. The first concerns itself with graph theory and yields the notion of the multiple relations of nodes;[30] the second is generated by social anthropologists and sociologists through the study of small groups (kinship units and social capital);[31] the third focuses on the study of organizations and firms, including multinational corporations, and by extension, society in general;[32] the fourth privileges the digital linkages that provide an infrastructure for the rise of virtual communities, digital diasporas, and cyberspatial forums;[33] and more recently, in a fifth approach, the concept has been applied to various forms of governance with or without government to account for the novelties brought about by the globalization process and information technology.[34]

Network governance has been studied in a variety of ways: in terms of how networks have emerged after a period of gestation; in terms of tacit interventions to change the modus operandi from hierarchy to network; in terms of linking separate entities into a network to make the operation more efficient; in terms of the construction of a network from scratch as an ad hoc structure, in contrast to a permanent structure; as cohabitating with a hierarchical structure; and as a chain of nodes operating under different legal regimes and maintaining different forms of relationships with one another.[35] From an analytic standpoint, various concepts that stress specific aspects of network governance have been used in the political science literature, such as "governance without a government,"[36] "multi-level governance,"[37] "co-management" or "co-governance,"[38] "metagovernance,"[39] and "governmentality."[40] Although network governance is seen as a post-bureaucratic model of organization, the assumptions underwriting this view have been found to be debatable.[41]

In all of these studies, the building blocks of the network approach have been the nodes (actors or units) and the relations (connections or bonds) that tie these entities together to form a social structure.[42] The assumption here is that understanding the nature of these relationships provides a way to decipher the behavior of each node and that of the entire matrix. In other words, and to be brief, the study of networks has emphasized their nature, function, deployment, the various types of relationships among nodes, the unequal importance of units, and even the underground structures developed and materialized in so-called dark networks, which undertake criminal or illegal activities.[43] However, such a methodological posture does not account for elements outside the network and their relations to the network. To correct this, I find it necessary to develop the concept of trans-global network governance as a more appropriate way to define the form of governmentality used by these scriptural religions.

The literature on "governance without a government," "network governance," "multilevel governance," and "metagovernance" has focused on the shift from a centralized form of government to a more distributed cellular form and on attempts to make sense of this evolution and of the everyday working of the process in both virtual and real time and space.[44] It does so by looking at the transformation that has happened, how it has affected the old form of governance, and how the new network form of governance is practiced. Analysts have found that in this process of the transformation of government, nongovernmental actors such as grassroots leaders, movements, and organizations can now influence state policies, international organizations, and policy outcomes. In addition, the private sector is now involved in setting new agendas for the state due to the central role of markets. In this analysis, however, I will focus less on the historical evolution of network governance and more on how a faith governs itself globally with or without a central administrative structure of governance to provide leadership to its multinational network of nodes.

Taking stock of previous findings, what follows unveils some characteristics of the global governance of Judaism, Islam, and Catholicism and advances the debate one step further, from network governance to transglobal network governance. Network governance speaks to relations between nodes, whereas the transglobal aspect refers to the spread of nodes beyond the boundaries of states and nations.

THE TRANSGLOBALIZATION OF NETWORK GOVERNANCE

Global governance evolves out of a new programming of society as a consequence of the construction of modernity as manifested in the creation of the nation-state in the eighteenth century.[45] The modern form of the state places global governance in a new international context. Although the three faiths existed before modernity, they were compelled to adjust to the requirements of the modern environment. As I just noted, the goal of this book is not to explain the genealogy of these various forms of global governance as influenced by national histories (the reunification of Italy, the centrality of the House of Saud in the Kingdom of Saudi Arabia, the destruction of the Temple of Jerusalem in 70 C.E., and the reclaiming of the territory of Israel and its transformation into a Jewish state, etc.). I recognize that remnants of these forms of governance persist in the new structures and that the shift from the old to the new is in itself interesting, but this is not the focus of this research.

In reference to Judaism, Catholicism, and Islam, the governance of each national node indirectly implies its participation in the governance of the global network, because one cannot be totally separated from the other, despite their different scales of intervention. In this view, the local site must be seen as a niche where the local node and the rest of the global network crisscross, interpenetrate, and cross-pollinate each other. Through this relationship, the governance of one (the local node) imbricates the governance of the other (the network). This explains why change in the global network may originate from the intervention of any node in the circuit.

Furthermore, it is important to clarify at the onset the meaning of network governance and to spell out some of its characteristic properties. Network governance can exist with or without government. For example, when the headquarters of a church outsources some of its activities, such as the Catholic Near East Welfare Association, it governs by facilitating and coordinating a network so that it can operate effectively. One version of network governance entails the existence of a relatively immutable core that connects and directs groups of people dispersed throughout the global landscape.[46] It is a network with a centralized commanding, operating, decision-making, and controlling bureaucracy. In this context, the headquarters gives itself a worldwide mandate to govern a network of satellite churches in order to achieve its global mission. The voluntary construction of such a network allows church leaders to maintain control, allocate tasks, coordinate output, expand the membership, and protect the faith. However, this view of network governance represents just one model: network governance with a government, as espoused by the Catholic Church, with its corporate headquarters in Vatican City.

But network governance also exists in a context where no node is recognized by the faithful as the center of authority affecting the activities of the entire membership. Each religious association develops its own governance structure and links with other groups to resolve common problems facing the community at large. In other words, each governs its own transnational or local segment and is responsible for harmonizing itself with others as possible allies for future ventures. The linkage with other groups may be done in a formal, contractual fashion or carried out informally by any members. Participation in such collective undertakings is done foremost not simply for the governance of their respective associations, but also and more precisely for the well-being of their homeland or diasporic communities and the practice of their collective faith. This view of network governance represents another model: network governance without a government, as espoused by Judaism.

In that model, attachment to an ancestral homeland, to the diaspora, and to the faith is the conditions that influence the form that network governance may take. Whereas units have developed different traditions of practice, they share the same putative ethnic background and faith in Judaism. In contrast, in the distributed cellular network governance model, each node has its own distinct national identity. Here again, each group governs itself within the parameters of its traditions. However, if any unit faces difficulty, groups with different traditions collaborate for the purpose of using common holy sites and protecting the community of believers. For example, once a year foreign ministers from Muslim countries, irrespective of their Sunni or Shiite backgrounds, meet to discuss the logistics for the hajj, the pilgrimage to Mecca. This represents distributed cellular network governance without a government, as espoused by Islam.

Transglobal network governance is not simply the public administration of a transnational organization, but it is also management of religious and moral values.[47] Religious governance gives a much larger scope to the administrative process, because it is seen as reflecting the values of the institution and therefore benefits not only its members, but also outsiders who may be attracted to join the religious institution. In a sense, religious governance is in itself a form of proselytism. Despite this common feature, the three faiths have developed and used different management paradigms to govern their networks of sites, paradigms that reflect their modes of organization and the unique religious values they want to sustain and propagate.

The network governance of global religions cannot be seen simply and exclusively in religious terms because of secular national and international pressures that influence its path. Global religions constantly converse with global society in an attempt to influence and change the global political and economic environment so that the faith can be practiced in a supportive milieu. Transglobal network governance means providing leadership to the networks of sites, protecting the borders of the networks (relations with other faiths and with nonreligious entities), and advancing the faith's mission outside its borders. Such a transglobal religious formation, which prepares to fight for those it controls, locates itself in and above the states it operates in, and eagerly seeks to expand its sphere of influence.

Global networked governance is multidimensional, because it comprises formal, informal, digital, segmental, and shadow networks. Formal networks are the focus of most studies, because they encompass the visible units of the global governance structure. Informal networks occur at the level of interaction between and among independent units, which either remain independent from the formal governance structure or constitute ad hoc organizations that informally discuss sensitive issues and gain access to hidden information. Digital networks comprise the webs of relations developed in virtuality that enhance performances in reality. Sectoral networks are relationships that engage some units, but not others, and shadow networks are underground webs of relationships and "dark networks" that are policed by the formal structure of governance to prevent them from negatively affecting the religious organization.

Furthermore, global governance requires the capacity to establish a regulatory system that addresses issues of compliance. How does a faith resolve conflicts if it is organized according to the governance without government model? In a hierarchical system, the central leadership may decide the case one way or another. In the absence of a central authority, the network does the arbitration. For example, Catholicism implements regulation through the nuncio system, through the annual report to Rome by dioceses, through the tribunal procedure summoned by the Vatican, through transfers of clergy to other dioceses, through compliance with canon law, and, in extreme cases, by excommunication. Judaism and Islam implement compliance through the requirements and application of halakhic or shariah law. These procedures are applied internally to those who belong to specific religious associations, because a religious movement does not have jurisdiction over other religious movements (e.g., within Catholicism, the Roman Catholic Church does not have canonical jurisdiction over the Eastern Orthodox Church, and vice versa).

The frontiers of global religions are key aspects that shed light on the internal properties of this form of network governance.[48] One must distinguish between different forms of frontier, because each differently affects the form of governance with or without a government. Institutional frontiers establish the boundaries of the religion; doctrinal frontiers spell out the religious identity and the beliefs of the religion; behavioral frontiers express its moral outlook and the lifestyle a religion expects from its practitioners; foundational frontiers characterize its specificity vis-à-vis other religions; administrative frontiers reflect the values of the religion that permeate its mode of operation; and the infrastructural frontiers dispel the ways in which the religion acquires and disposes of its physical properties.

Although this description of network governance may seem extensive, there are key aspects of this network form of governance that have not yet been mentioned. Among these are "a trend for a wider range of participants to be seen as legitimate members of the decision-making process" and the need for religious leaders "to be open to learning in different ways, and to draw in resources from a range of sources," as Gerry Stoker puts it.[49] Some networks are an "instrument for cooperation," whereas others are a mechanism for domination, and still others are a tool for self-regulation.[50] Some salient characteristics of this form of governance can be identified as follows: rotation (any node can exert leadership of the network); relative equality (no node serves permanently as the leader of the pack and people from other nodes can be recruited to staff a node); mutual dependence and responsibility (no node is an island into itself, and every node is responsible for the welfare of others); the network governs itself, and each node contributes to the successful performance of the network and its reproduction; the network displays a malleable and variable architecture, depending on circumstances; and nodes do not behave in identical ways, because they express themselves in different state environments and demonstrate differences of culture, history, and traditions.

THE GLOBAL MOBILITY OF NETWORK GOVERNANCE

Global network governance is a mobile structure that reflects the movement of people from one place to another, the movement of monetary donations from one country to another, and the movement of information from a local site to any node of the global landscape. This global mobility of network governance is more precisely detected in the overseas transfers or appointments of clergy, overseas specialized training of students as a preselection for their assignments to global leadership positions, the transnational movements of large groups of people to attend the hajj in Mecca, the international sharing of resources through the management of assets, multinational collaboration to ensure the security of places of worship, global policing of the boundaries of the faith, and global distribution and commercialization of church-sanctioned dietary products. Even when these activities are carried out locally, they contain a global component, because each node is part of a transnational network of sites.

The global mobility of religious bodies is a characteristic function that falls under the rubric of this category of network governance.[51] The annual Muslim pilgrimage to Mecca has created logistical management problems for the Kingdom of Saudi Arabia. As a result, an international governance structure has been set in place to smooth the process of journeying to Saudi Arabia by air, land, and sea and thereby to prevent housing crises, food shortages, traffic jams, health hazards, and security nightmares during the pilgrimage period. At the annual meeting of the foreign ministers of Muslim countries, participants fix the quota of pilgrims per country and discuss other logistical matters.

Network governance in this particular case has led to three areas of global practice: the creation of government ministries whose main function is to take charge of the pilgrimage, the transformation of the infrastructure of the country to meet the needs of pilgrims, and the establishment and management of a quota system that assigns a specific number of visas to each country so as to control the flow of visitors per year.

The cross-border transfer of religious leaders, which is a form of international mobility of religious bodies, is a prerogative of any global religion in its capacity to govern its transnational network of sites and practitioners. Transfers can be permanent, transitional, or occasional. A permanent transfer is undertaken for the sake of appointing clergy to permanent posts abroad. The pope sometimes appoints a Vatican bishop to a post outside of Italy as head of an archdiocese. This is done to ensure that dioceses without their own bishops can be provided with one from elsewhere or to provide pastoral experience to a priest or bishop who has served as a bureaucrat at the Curia. This is common practice at the national level when a priest or bishop is transferred from one region of the country to another in order to care for a parish or a diocese.[52] The same phenomenon happens when missionaries are sent from one country to another to take care of an operation such as a school, a church, or the implantation of a religious order. Global transfers can also be transitional, as in the case of the appointment of nuncios. Nuncios represent the Vatican in a specific country and are usually recalled after they complete their term of office to be reappointed elsewhere or to be sent back to the Curia. In some cases, they decide or are asked to return to parish life in their homeland. A transfer is said to be occasional when it concerns transnational mobility for a short period of time, as in the case of papal envoys who perform specific diplomatic missions. These are delicate missions that require either a senior Vatican diplomat or a bureaucrat with special powers conferred by the pope for negotiations on his behalf.

The global mobility of actors in the three faiths—Judaism, Catholicism, and Islam—is regulated through network governance to maintain order, prevent abuses, and ensure that pressing needs are met. In Catholicism, the governance of this transnational mobility is done not only in the case of appointments either initiated by the Vatican or requested by the local bishop, but also in the context of the mobility of religious leaders to attend bishops' meetings in other countries (synods, councils, funerals, elections of the pope, and requests to meet with the pope for an emergency matter, etc.). The Vatican monitors extraterritorial mobility not only to ensure that bishops remain at the post, but also to prevent the misuse of church funds.

The network governance of the mobility of religious leaders in Judaism follows a different logic. In Judaism, rabbis from one country can be appointed to serve a congregation in another country on the basis of contractual arrangements between the two parties. For example, a rabbi who graduated from an Israeli seminary may be offered a position in a foreign country to lead a diasporic congregation. In this instance, where the needs of one node are met by the availability of leaders in other nodes, there is no central authority to monitor these movements. A rabbi from the Orthodox tradition may be selected for a post at a Conservative synagogue.

The network governance of Wahhabi Islam is also engaged in the proper conduct of the mobility of its religious leaders. This concerns most particularly the management of the mobility of those who may be assigned to lead the faithful in foreign countries and those who serve as itinerant teachers of Arabic, invited guest preachers abroad, or participants in academic meetings for jurisprudence scholars (shariah law). This governance materializes through proper training provided to future leaders, salary payments, and the selection of appropriate volunteers to represent the faith in foreign lands.

The principle of the global transfer system shows that the religion as a global institution supersedes nation-states because it has tentacles attached to most of them, transcends the vicissitudes of local arrangements, and deploys network governance in a way that affects local branches and the mobility of practitioners.

The network governance of the mobility of religious bodies is also evident in the training of seminary students for global leadership positions. Every year, Muslim students, scholars, or imams from all over the world come to the Kingdom of Saudi Arabia to immerse themselves in the study of Islam at the Islamic University of Medina. The Saudi government provides most of them with scholarships to help defray expenses with the understanding that when they return home, they will be able to share their knowledge with their respective communities.

Each of these global religions necessarily engages in global training for the transmission of its doctrine, its social reproduction, the professional management of its property, and its global expansion. For example, the majority of Catholic priests train in their homeland seminaries staffed by instructors who speak the language, who are socialized in the culture, who are knowledgeable about the people, and who are familiar with the national school and university system. Of course, these seminarians are taught courses on the history of the Church and canon law. In contrast, professional global cadres are those individuals who have been selected to attend Roman universities and are de facto potential bishops, nuncios, Catholic seminary or university professors, or even potential bureaucrats at the Curia. They are brought to Rome for special training, to familiarize themselves with the bureaucracy of the Church, and to share that understanding with the people in their own countries. They are trained to become the best allies of the Vatican in their homelands.

Global training in Rome includes exposure, under controlled conditions, to what the Church holds to be the orthodox interpretation of its teaching and doctrine. Those whom the Roman universities or seminaries invite to teach are usually North American and Western European theologians who are known to follow the pro-Vatican orthodox path. Some are priests and others are members of religious orders. These Roman universities or theology schools such as the Pontifical Lateran University, the Pontifical Gregorian University (Jesuits), the Pontifical University of St. Thomas Aquinas (Dominicans), or the Pontifical University Antonianum Atheneum (Franciscans) belong for the most part to religious orders or are pontifical seminaries that specific countries have established in or near the Vatican (the North American College, the Irish College, the Canadian College, the Belgian College, the College of St. Jerome of the Croatians, the Polish College, the Philippine College, the Ethiopian College, the Brazilian College, and so on).

For the same reasons, Judaism follows a similar tradition of training rabbinical students in Jerusalem, whether for a semester or for a year. These students pay for the training with the help of their families or receive scholarships from either donors or the rabbinical school. Some are able to make arrangements with their local universities for the transfer of their credits. The training here is for global leadership positions, and some are able to land jobs in synagogues around the globe.

Another aspect of the mobility of network governance is in the way in which global religions manage their global assets, most notably their property. Building institutions, paying staff members, educating religious leaders, managing real estate, defraying transportation costs, and printing books—prayer books, hymns, Bibles, or Korans—for free distribution all cost money, despite the benefits of proselytism and expansion. Local nodes are seldom able to pay for all of these expenses, which is why the network provides the necessary resources to complete these ventures. Local physical entities depend on the network not only during the initial phase of operation, but also throughout their life cycle, when they continue to rely on the network for resources and guidance. In theory, synagogues belong to local congregations. In practice, a large portion of the money to operate some synagogues comes from donors outside the area, including national or international Jewish organizations. In the case of Islam, the Kingdom of Saudi Arabia has contributed to the building of mosques in various countries around the world. In this same vein, the Holy See and National Catholic Organizations in Western Europe and North America sometimes contribute money and goods to local churches in poor countries, not simply to erect buildings, but also to provide them with relief aid.

INFORMATION TECHNOLOGY AND NETWORK GOVERNANCE

The efficiency of network governance often materializes through the use of IT—information technology—which enhances communication and exchange.[53] IT has strengthened the globalization of religious governance by allowing leaders to communicate with each other, by making documents available online, and by facilitating virtual transactions for those who want to make financial contributions to the religious institution. In the past, leaders and their representatives communicated with each other by mail, via the telephone, and by facsimiles. More recently, the Internet and e-mail have been added to the communications system.

The impact of IT on network governance can be detected in several ways. For some institutions, IT may add a virtual dimension to what existed before, whereas for others it can transform aspects of the traditional mode of operation. It can also link nodes in the same global network to each other and enable their interaction. In some cases, IT even links disparate units and turns them into operative nodes of a global network, which remakes them and provides them with a new identity.[54]

Network governance also includes the digital governance of the network. This involves the deployment of the infrastructure of communication (communications equipment, computers, and other digital gadgets) and of the digital network, with its multiple communicational hubs, as well as the use of the network to divulge and receive information digitally. To be brief and precise, digital governance involves the management of the digital infrastructure and the digital use of the platform to ensure the greater moral benefit to the community of the faithful.[55]

The digital infrastructure is not a given and must be constructed to meet the specific needs of a religious institution and network. It must be developed in harmony with the teaching of the religious organization, the geographical spread of the network, the availability of resources, and the extent to which it is useful to the faithful, depending on the access, literacy, and expectations of users. This explains the differences in digital infrastructures.

As in the case of e-government, the digital governance of global Judaism, Islam, and Catholicism has been accomplished in two prominent ways. Each religious institution now has its own website, which provides information concerning doctrinal issues, church celebrations, homilies or speeches by leaders, the position of the religious organization on specific secular issues, the kinds of services it provides, and its ongoing projects. This static information is made available on the Internet so that people can access it at will. The second dimension is the use of the Internet for the exchange of information and for transactions. An individual can e-mail a request for information and make a pledge online to provide a donation to the faith, or the faithful can use a bulletin board to exchange information and a chat room to discuss matters related to their religious life. Through these practices, the digital realm enhances and contributes to the traditional forms of communication among the faithful.[56]

THE LOGIC OF MULTINATIONAL GOVERNANCE

Religious network governance has a global logic because each node is concerned about the governance of the network and the governance of the unit. On the one hand, the governance of a node allows the node to maintain its religious identity while it contributes to the governance of the network, and on the other hand, the governance of the network allows nodes to rely on each other in times of crisis and thereby define the identity of the network. For some, governing the network means managing a collective project and infrastructure, but not necessarily the content of the faith, whereas for others, it means managing the faith; still others may engage in both enterprises.

The global logic of network governance cannot be studied only in reference to central management. Because each node is part of the global network, each node must operate in a global environment. For example, knowledge may be generated in a node, but approved and consumed by the network. In addition, help may be given by one node to other nodes in the network, or a position in one node may be filled by someone from another node. Although members may live in different countries, they share the same transnational religious space, hence the metaphor of the *ummah* or *ecclesia*, which treats practitioners of the faith as brothers and sisters of the same religious family.

Global governance of the network entails both a multicultural makeup of the leadership and multicultural representation. In network governance without a government, as seen in the case of global Judaism, a multitude of nodes with equal status constitutes the transnational circuit. Diversity of leadership is achieved in the interaction of these nodes with each other. In the other case of network governance without a government, in global Islam, the nodes tend to perform as national units, because secular politics strongly influences their transnational interaction. Network governance in this instance is perturbed by the strength, power, and wealth of any national node. In contrast, in network governance with a government, as seen in the case of Catholicism, the leadership node or the Holy See that is endowed to govern the whole is multicultural, and the Vatican calls upon bishops and cardinals from diverse national backgrounds to occupy key positions in its governance. For example, during John Paul II's papacy, Benin's Cardinal Bernardin Gantin served as prefect of the Congregation for Bishops, Vietnam's Cardinal Nguyen Van Thuan as president of the Pontifical Council for Justice and Peace, and Nigeria's Cardinal Francis Arinze as prefect of the Congregation of Divine Worship. Through the conclave, which is an electoral process, the pope is elected.[57] The multinational composition of the Electoral College that participates in the conclave ensures transparency in the contribution of the network of nodes to the process. Because the Roman pontiff has the power to select the members of his cabinet, however, the cardinals with Curia experience may have an opportunity and a strategic advantage in vying for the office of the papacy when it becomes vacant, as we have seen in the cases of Pius XII (formerly Secretary of State), Paul VI (formerly Pro-Secretary of State), and Benedict XVI (formerly Prefect of the Congregation for the Doctrine of Faith and Dean of the College of Cardinals) (Table 1.1).

Table 1.1 Papal Election and the Curia Connection: 1800–2005

Cardinals	Nationality	Previous Curia Position	Position During Conclave	Papacy
Cardinal Joseph Ratzinger (1977–2005)	Germany	None	Prefect of the Congregation for the Doctrine of the Faith (1981–2005)	Benedict XVI (2005–. . .)
Cardinal Karol Wojtyła (1967–1978)	Poland	None	Archbishop of Krakow, Poland (1964–1978)	John Paul II (1978–2005)
Cardinal Albino Luciani (1973–1978)	Italy	None	Patriarch of Venice, Italy (1969–1978)	John Paul I (Aug. 1978–Sept. 1978)
Cardinal Giovanni Battista Montini (1958–1963)	Italy	Pro-Secretary of State (1952–1954)	Archbishop of Milan, Italy (1954–1963)	Paul VI (1963–1978)
Cardinal Angelo Giuseppe Roncalli (1953–1958)	Italy	Apostolic Nuncio of France (1944–1953)	Patriarch of Venice, Italy (1953–1958)	John XXIII (1958–1963)
Cardinal Eugenio Pacelli (1929–1939)	Italy	Apostolic Nuncio of Prussia (1926–1929)	Secretary of State (1930–1939)	Pius XII (1939–1958)
Cardinal Achille Ratti (1921–1922)	Italy	Prefect of the Vatican Apostolic Library (1914–1919)	Archbishop of Milan, Italy (1921–1922)	Pius XI (1922–1939)
Cardinal Giacomo Della Chiesa (1914–1914)	Italy	Undersecretary of State (1901–1907)	Archbishop of Bologna, Italy (1907–1914)	Benedict XV (1914–1922)

Continued

Table 1.1 Continued

Cardinals	Nationality	Previous Curia Position	Position During Conclave	Papacy
Cardinal Giuseppe Sarto (1893–1903)	Italy	None	Patriarch of Venice, Italy (1893–1903)	Pius X (1903–1914)
Cardinal Gioacchino Pecci (1853–1878)	Italy	Apostolic Nuncio of Belgium (1843–1846)	Camerlengo of the Holy Roman Church of the Apostolic Camera (1877–1878)	Leo XIII (1878–1903)
Cardinal Giovanni Maria Mastai Ferretti (1840–1846)	Italy	None	Archbishop of Imola, Italy (1832–1846)	Pius IX (1846–1878)
Cardinal Bartolomeo Alberto Cappellari (1826–1831)	Italy	None	Prefect of the Congregation for the Evangelization of Peoples (1826–1831)	Gregory XVI (1831–1846)
Cardinal Francesco Saverio Castiglioni (1816–1829)	Italy	None	Prefect of the Congregation of the Index (1821–1829)	Pius VIII (1829–1830)
Cardinal Annibale della Genga (1816–1823)	Italy	Apostolic Nuncio of Cologne (1794–1802)	Archpriest of the Papal Basilica of St. Mary Major (1821–1823)	Leo XII (1823–1829)
Cardinal Barnaba Chiaramonti (1785–1800)	Italy	None	Bishop of Imola, Italy (1785–1800)	Pius VII (1800–1823)

SCALES OF ANALYSIS

In what follows, this study analyzes the network organizations of these three religious institutions by using different scales to penetrate and explain their modes of global governance.[58] It focuses on the total network in the case of Judaism to show how network governance works in this global religious setting. It singles out the Roman Catholic Church for the study of its global governance logic because of the visible corporate dimension of its leadership structure. In doing so, it illuminates how a dominant node functions in network governance. It then examines Wahhabi Islam as one of the units of global Islam to illuminate the functioning of a competitive node in a distributed cellular form of global governance. These different entry points provide clues to a multitude of mechanisms that yield information on how each religion proceeds in its global governance operations. In this kind of analysis, one penetrates the network via any available nodes, because each node provides useful insights to understand the functioning of the whole.

Methodologically, the focus on the governance of the network allows us to study it as a spatially dispersed entity accomplishing tasks similar to those of a corporate headquarters. The focus on nodes provides insights into their contents, roles, interactions, and modes of relations with the network. The focus on relations between nodes, in turn, explains how they affect the functioning of the network or how each node is influenced by the network. For example, the network may reach a consensus on a specific dimension of the faith, but a node may attempt to influence the network and may achieve success in so doing or may simply take a dormant position and reactivate its participation at a later period.

In essence, the mode of analysis that I follow in this investigation indicates a shift from a state-centric to a global methodology.[59] I identify the global arena instead of the nation-state parameters as constituting the key platform of performances of cosmopolitan actors or distributed cellular entities.

The central project of this research is the analysis of the deployment, behavior, and performance of these global religions through the kaleidoscopic lens of network governance. This form of globalization has been in practice for many years and has affected all of the nation-states in one form or another. Yet very little is known about the ways in which global religions manage the widespread nature of their organizations and the dispersed tentacles that constitute their membership. The lack of knowledge in this area handicaps any effort to understand the transnational integration of the global community of nations and the continuing search for signposts in the global inscription of multicultural metropolises.

2 Global Judaism and Cosmonational Network Governance

Israel is the magnetic pole whose lines of force are crucial to the ability of the Jews to orient themselves. . . . Israel is no Vatican City: Israel's machinery of government is no Curia; its ambassadors are not Papal nuncios, let alone legates.

—David Vital, The Future of Jews[1]

What has emerged is a matrix of institutions and individuals linked to a unique communications network; a set of interacting institutions which, while preserving their own structural integrity and filling their own functional roles, are informed by shared patterns of culture, activated by a shared system of organization and governed by shared leadership cadres. . . . In some communities, the network is connected through a common center which serves as the major channel for communication. In others, the network forms a matrix without any center, with the lines of communication crisscrossing in all directions. . . . There is rarely a fixed division of authority and influence but, rather, one that varies from time to time and usually from issue to issue, with different elements in the matrix taking on different loads at different times and relative to different issues.

—Daniel J. Elazar, Land, State, and Diaspora in the History of the Jewish Polity[2]

Judaism, like Catholicism and Islam, is a global religion not simply because of its universalism as a faith, welcoming both Jews and Gentiles to its fold and projecting a message of salvation and an exemplary code of moral conduct, but also because of the geographical dispersion of its adherents as a result of immigration and diasporization after the destruction of the Second Temple in 70 of the Common Era. The faith that was once confined to the Middle East and, to a lesser extent, the Mediterranean region has spread throughout the world because of successive migrations due to trade, persecutions, expulsions, family reunifications, colonization, searches for employment, and the Holocaust. As a result of these worldwide population movements, and despite the adverse circumstances that religious Jewish immigrants have sometimes encountered, wherever they have resettled, they have kept the faith and sustained their longing for their eventual physical and eschatological return to the land of Israel.

This chapter will not delve into the myriad ways in which transnational Jewish networks operated prior to the end of World War II, even though such an inquiry could explain the reproduction of Judaism in diaspora over time and the relations between the diasporic sites and Palestine, including the old Jewish Quarter in Jerusalem. This kind of genealogical approach to network formation, deployment, and collapse, although interesting and instructive in and of itself, is not the focus of our investigation here. Rather, this chapter explains the mechanisms of network governance of global Judaism from the independence of the State of Israel in 1948 to the present. It is not an exhaustive study of the phenomenon, but it unveils some essential elements of how network governance is deployed pertaining to relations between sites, the training and mobility of rabbis and religious teachers, the protection of synagogues, the global infrastructure of dietary practices, and the management of the sharing of financial resources with other sites in the global network.

The study assumes that in order to understand local religious practices and the behaviors of their practitioners, one must conceptualize the operation of the larger networks of relations. In that light, I argue that network governance plays a key role in this global dynamic of the faithful. The ebb and flow in this global network is unveiled not simply in regard to questions of orthodoxy or orthopraxis, but also in the area of infrastructural linkages—vertical relations for liturgical coordination and defense purposes and horizontal relations in order to help communities in distress. The focus on global governance helps us to explain the modalities of local governance and the linkages between various sites and elucidates how the network self-governs in the absence of a corporate control center similar to the Vatican that rules on practices of noncompliance, and without a formal legal code that establishes the criteria for relations between nodes, as exists in the Catholic Church through the use of canon law. Network governance without a government is the form of governmentality that best describes global Judaism as practiced locally by Jews in Israel and the diaspora. The transnational religious leadership accounts for Judaism's intrinsic coherence as a collective faith and its survival over time.

NETWORK GOVERNANCE

The concept of network governance as applied to global Judaism makes sense only within specific parameters, because the religion operates in a unique way that distinguishes it from the other monotheistic faiths. In global Judaism, the internal networks are part of the daily operation of the religious organization, whereas the external networks, which concern the relations of the religion with groups of other faiths and civil society in general,[3] are not always active and may remain dormant until they are either reactivated or reengineered. No existing theory of network analysis conveniently fits the rhizome-like deployment of global Judaism. Thus, it is important to set forth several different methodological considerations in order to shed light on specific aspects of the functioning of networked Judaism.

In Michel Foucault's approach to the "conduct of conduct," governmentality prioritizes agency over structure and conceives of governance more in terms of disciplinary technologies for the production of subjectivities and identities than in terms of strategies to achieve common goals for the citizenry. It conveys concerns over meanings, personal conduct, and the art of governing. Rules, codes, and disciplinary practices are supposed to constrain and shape individual behaviors and are how society goes about producing the individualization and naturalization of subjects.[4] The strength of his formulation is that it unveils linkages between the personal and the political and between culture and power.[5] However, the limits of Foucault's approach reside mostly in its nationalist methodology. He frames his concepts within the context and boundaries of the nation-state. This nation-state framework becomes less useful once we place the object of study in a global context, where identity becomes less stable, more fluid, and prone to take multiple forms. Under these circumstances, it becomes more difficult to pinpoint any single institution as responsible for the processing of identities. For the broader analysis in this book, Foucault's concept of the "circulation of power" must be reformulated in terms of the mobility of governance, a mobility in which circulation does not mean a rupture with the network, but rather the repositioning of agents and entities within it.[6]

In analyzing network governance, I therefore deliberately choose to shift attention away from an analysis of institutions to an analysis of the *procedures of practice*, focusing my attention not on what network governance is or what it does, but rather on *how* it does what it does. This allows one to understand the role of the ensemble in the production of individual practices, but also how individual practices may affect the entire circuit.[7] This strategy shows how malleable the network can be, how networks are made, unmade, and remade, and how there are multiple opportunities for individual entities to reposition themselves within them. By procedures of practice, I mean the techniques, interventions, strategies, performances, and other ways in which the network and each node undertake the business of governing and administering the ensemble.

In Judaism, network governance is not government by a leader or a group, but rather government by different groups in different sites at different times with the tacit consent of the global circuit. As a former dean of a Jewish school network on the East Coast of the United States who is now retired in Jerusalem pointed out, there is no formal governance of Judaism, but rather a loose network operating on the basis of the strengths of each community. The linkage, according to his observations, is both ethnic and religious. Each group contributes independently or in conjunction with others to the survival and social reproduction of the faith and the ethnic identity.[8] Harmony and the pursuit of the collective interest or the common good are what make this form of governance sustainable. The network in this specific context of global governance is the embodiment of the message.

Seen in this light, governing the local is an intrinsic component to the governance of the global. For example, in the scheme of the relations between a nation-state and its secular diaspora, governance can be constructed as an addendum, because the state can govern itself without paying attention to the diaspora, albeit perhaps to its peril. In contrast, in the case of the governance of a global religion, the relations of the global to the local are intrinsic to the governing process, because they redefine the local not simply as a site, but rather as a node of a transnational network.

The network governance model of Judaism inescapably leads us to examine the organization of the faith in relation to various sites where its membership is located. Not all the nodes in a network are always active or have the same types of relations that link its diverse membership. In such situations, multiple forms of the governance of nodes and their linkages to the network constitute the backbone of network governance. In such situations, network governance is not the governance of homogeneous entities, but rather the result of the "conduct of conduct" in heterogeneous nodes.

Although local membership largely influences the governance and administration of a local synagogue, this practice ensures that different congregations in different countries develop their own local traditions. In any case, each local synagogue hires the rabbi that best suits its needs. It must also be said that ethnic needs intertwine with religious needs so much that network governance of the faith does not concern only religious leaders, but also secular leaders who play a role in its performance through fundraising, donating money, staffing Jewish agencies, commercializing kosher foods, as well as through other means. In other words, the governance of the social infrastructure of the group cannot divorce itself from the governance of the faith, because the former sustains the performance of the latter.

Because no unit is officially in charge of the entire network, the position of sites varies over time. Because of persecution and the migration of leaders to other countries, certain sites become less prominent than others. In addition, a site that was not important may become prominent due to population increases and an influx of émigrés, as in the case of Jewish New York after World War II. As Harvey Goldberg puts it, paraphrasing Simon Dubnow (1860–1941), "new centers of Jewish life emerged in different parts of the globe . . . replacing those that had declined. Babylonia succeeded the Land of Israel, which in turn was succeeded by Spain and the Rhineland, and the latter two were succeeded by Eastern Europe."[9] One might add that the last center of Jewish life now has been succeeded by the State of Israel,[10] because since 1948, the European node has been marginalized to the advantage of Israel and the United States, and since the Six-Day War in 1967, Israel has become a central operational factor in the definition of Jewishness for religious Jews. The centrality of Israel has come about not only because it is the putative "homeland," but also because the faith is attached to the territory, including its holy sites where it took its early shape.

In this scheme of things, the mobility of network governance involves the changing strength or weakness of sites over time and shows how the prominence of a site may restructure the behavior of the network. Mobility can transform a satellite node into a central node, thereby gaining responsibilities vis-à-vis other sites and the network in general. What is of paramount importance here is not the fact that a node becomes prominent, but that the network now has a different architecture, which in turn affects other nodes in new ways.

In Judaism, no mechanism of network governance has developed to serve as a structure of command that interjects itself in all the nodes that populate its landscape. Consequently, to explain how global Judaism operates and reproduces itself over time, we need to look for not only a governing structure, but also for loose points of connection between the nodes. Nodes forming the aggregate sites that thus characterize the architecture of this network cannot maintain their identities if they disconnect from the whole that directly or indirectly feeds their separate existences.

As already noted, in this universe, global networks are composed of nodes with different weights and histories that contribute differently to the equilibrium and sustenance of the whole. Some examples of uneven nodes include: the Consistoire in France, the Kultesgemeinde in Central Europe, the United Congregational model in England, the Congregational model in the United States, and the Landsmannschaft association model in Latin America.[11]

In addition, these nodes form crisscrossing subnetworks with the network of global Judaism. For example, Conservative Judaism has a global network that is different from those of Orthodox or Reform Judaism. These subnetworks are both parallel and intersecting—parallel because they have their own memberships, theological orientations, and ritual practices, but intersecting because their memberships participate in joint civil undertakings to uphold common group interests, including the security of the State of Israel, a concern shared with nonpracticing Jews as well.

MULTIPLE GLOBALITIES

Network governance in the case of Judaism thus involves multiple globalities in diverse global subnetworks that perform different functions for their constituencies, have their own agendas, and are imbued with their own distinct logics or rationales. They normally and routinely operate on a global scale to fulfill their own missions while sporadically connecting with other networks to serve as sister organizations outside the parameters of their everyday operations. The Lubavitch Chabad movement, with its headquarters in Brooklyn's Crown Heights, for example, has established well-known outposts in Europe, Israel, and Latin America.[12] It is a global undertaking with global networks of congregations, rabbis, and institutions. The landscape it covers is not identical to the sphere of influence of other religious Jewish movements. Network governance in this case has to involve both the activities of these diverse subnetworks and the extralocal connections between them. In the case of global Judaism, this includes the governance of religious leadership networks, security networks, charitable action networks, dietary infrastructure networks, and synagogal networks. Members of these networks globally connect their organizations to other institutions or other sites. Structurally, where two or more such subnetworks meet and crisscross each other, they form nodes in the larger network.

The governance structure constantly redeploys itself by deploying actors at the intersection of such networks. For example, in Berlin, the security network for the protection of synagogues includes Israeli guards (young adults who have completed their military service in Israel), diasporic Jewish personnel, professional security-agency personnel, German policemen, and Israeli secret agents. These layers of security, which become visible during Jewish holy days and celebrations, are part of a necessary global infrastructure that sustains the practice of Judaism in Germany.[13] Similar security apparatuses can be found in France and other countries inside the European Union.

The network governance of these multiple globalities expresses itself through the religious movements within global Judaism, the global infrastructure that sustains the faith in its global manifestation, the global community of practitioners that reproduces it over time, and the global network of relations that serves as its engine.

THE LOCAL SYNAGOGUE AS A GLOBAL NODE

From a global perspective, Judaism can be divided into three main categories: *synagogal Judaism*, in which the congregation is the driving force of the faith-based and autonomous community; *associational Judaism*, in which synagogues are offshoots, branches, or units of the Conservative, Orthodox, or Reform movements; and *state Judaism*, in which the state—as happens in the case of Israel—influences the shape of the hierarchical and organizational structure of the religious leadership, financially contributes to the religious organization, and claims responsibility for the security of synagogues.

The local synagogue thus may be part of a recognized national structure or may be located outside such parameters. For example, some French synagogues fall under the aegis of the Consistoire, created by Napoleon in an effort to bring Jews under the umbrella of the French state, whereas others do not. This inconsistency creates different practices in terms of local governance. Those that do not belong to the Consistoire network operate either as autonomous communities, independently in terms of the management of their finances, or as offshoots, branches, or units of the Conservative, Orthodox, or Reform movements. All receive donations from their congregations as the principal source of revenue for the maintenance of their operation. However, synagogues affiliated with the Consistoire receive financial aid from the global network when they experience hardships. Not only does the Consistoire at times help make up the budget deficits of some synagogues, it also participates in planning for and contributes financially to the construction of new synagogues. Synagogues that are part of a religious movement undertake their activities in harmony with the specific religious subnetwork to which they belong, including the rebbe who is their spiritual leader, as in the case of the Hasidim. In a time of crisis, independent synagogues may seek or accept aid from any Jewish organization or agency. Although the relations of any synagogue with any Jewish agency may be dormant, they can be activated in a time of urgent need.

There is a major distinction between synagogues in countries where the government finances the construction and maintenance of synagogues and those in which the local congregation controls these aspects. In states like Israel that provide financial support, buildings tend to be less elaborate, and architecture is more modest and uniform in terms of appearance and decorum because all the congregations receive the same level of funding. In contrast, when the construction of a synagogue depends on the congregation, the building tends to reflect the poverty or wealth of the community or movement. Congregations use their place of worship to display their wealth and to compete with other communities.

This difference between state- and non-state-financed synagogues also appears in the expression of Jewish identity and the status of the rabbi. In North America, the synagogue to which a congregant belongs very often influences the performance of his or her Jewish identity. There, the Jew whose affiliation lies with an Orthodox synagogue pays dues to that synagogue in exchange for the rabbi's devotion to the synagogue and its congregation. In Israel, by contrast, the town rabbi serves in an elective state position with a monthly salary from the government. In fact, there are few rabbis in Israel who are paid full time to serve a specific synagogue. Many rabbis work on a part-time basis or simply acquire honorary assignments as part of the religious leadership team of a synagogue. In these cases, they provide services, such as answering the questions of the faithful, but they often do other things to make a living. This dynamic exists because the town rabbi is a government position, not a synagogue position. For example, the Jerusalem Great Synagogue has its own rabbi, plus an honorary rabbi and a cantor.

Every synagogue in North America is owned by the congregation that builds, manages, or uses it. No central national or international organization controls the process of their acquisition or the sale of these synagogues. The situation is similar in Israel; although there is a Chief Rabbinate in Israel, it does not own the synagogues. Therefore, in Israel, synagogues may be owned by the people who raise money to build them or by the congregation that constitutes their membership.

Some synagogues in the United States are affiliated with a larger body, such as the United Synagogues of America, which is a part of the Conservative movement, or its competitor, the Union of Orthodox Rabbis, which is the Orthodox alternative.[14] This is just an affiliation and, therefore, does not imply any financial connection or dependence. If a Conservative synagogue in the United States wants a rabbi, the local leadership may ask the United Synagogues to help find an appropriate applicant, such as a rabbinical student from their school. But such a rabbi cannot be imposed on the congregation if it does not want him or her. Certainly, the central body does not own such a satellite synagogue in any way.

However, even the synagogues that are affiliated with religious institutions, such as the Union of Orthodox Congregations or the Rabbinical Council of America, also maintain their autonomy and look after themselves. The umbrella organizations with which they associate may not necessarily help the membership in the everyday running of their operations, but, for example, may help with funding retirement insurance. The situation of Jewish schools is similar to that of the local synagogue. In the United States, there is a very wide network of Jewish schools. Although each student pays tuition, the congregants must raise the rest of the money. Here again, each town, city, or metropolitan area with a significant Jewish population has its own schools and does its own fundraising to manage its educational facilities. Each school is a separate entity that may rely on an umbrella organization to provide certain services, such as curriculum development and implementation, but not management or fundraising.

The fact that the local congregation owns the place of worship allows it much more freedom to organize itself as it sees fit. In contrast to the status of a congregation under the Catholic Church's policies, the congregation of a local synagogue can manifest its independence more freely and experiment with new forms of ritual that do not need approval from a central bureaucracy. Should the synagogue decide to secede from a religious movement, it can do so without the risk of a legal lien on the building, because the property belongs to the congregation, and not to an external entity.

The local synagogue has both a global and local vocation. It is a local institution because it caters its services to the local community in terms of the language used in the liturgy, the authority that governs its everyday operation, the membership that pays for its survival, the local problems it attempts to resolve, the location of the space that it occupies, and the relations it must maintain with the neighbors and the city that sets zoning regulations.

As part of a global network, in matters of management, finance, and the appointment of rabbis, synagogues thus possess a certain independence of action and much latitude in terms of their national affiliation to larger global structures. What therefore concerns us most here is how autonomous synagogues, shaped by the local culture of their environment and the history of settlement, are linked to each other in the maintenance of the faith. Because synagogues do not function by the imposition of governance from the top down, but by the exercise of discernment and power by the membership, the independent synagogue is the most visible formal link that locates a group in the global network of nodes.

A synagogue is a global and multicultural node because of the diverse national backgrounds of the membership, including those who have emigrated from various countries and those who happen to reside in the neighborhood. It is globalized by the attachment to, and continued contact of, some members with spiritual leaders in their homelands through phone calls, e-mails, and periodic visits, by their relations with secular institutions with global agendas, by their linkage with Israel for monetary and religious purposes, and sometimes by their structural positions in a larger religious movement. The synagogue is also global because overseas Jews may attend services when they visit family members in the neighborhood, and because of its relations with foreign Jewish communities that may request help in case of distress.

Another aspect of global governance that facilitates its deployment and sustenance is the way in which a synagogue recruits its membership locally. In any given synagogue, the Orthodox, Conservative, and Reform membership may mix when the place of worship is the only one available to the community in the area. The alternative for members of such a community is either to join this mixed assembly or simply revert to a privatization of practice. Migration and the intermingling of practitioners at the local level constitute a mechanism for the sustenance, reproduction, and globalization of the faith. The focus here is not on linkages between the leaders, but rather on the multicultural and plural composition of the local congregation. This complicates any theory of network governance that does not focus jointly on upper-, middle-, and lower-level types of global relations. Also, global governance can be seen not simply in terms of multiple extraterritorial relationships, but rather in terms of the global composition of localities.

Local synagogues thus play a double global role because they are linked to overseas entities, both formally and informally. In this sense, local synagogues contribute to the shape of the global and the global network contributes to the shape of the local. For example, in the course of an interview with a Jewish woman in Scheunenviertel, the site of the former Jewish ghetto in Berlin, I discovered that in addition to her participation in a Lubavitch synagogue, she flies to London once a year for spiritual consultation with a rebbe who became acquainted with her in Berlin and later transferred to London. Despite his relocation, she is still attached to his teaching and seeks him out for spiritual guidance. Here is an example of how a local group may intersect with the global through its leader or that transnational relations may occur through individual members' initiatives. For example, some maintain their membership in a local synagogue but also remain in contact with congregations in other countries. This double membership is one example of a linkage occurring through individual members' actions, not through a formal group arrangement.

Global Judaism thus manifests itself through a network of synagogues that are diverse, multiple, and autonomous because the congregation locally controls them. Some synagogues are aggregated as part of a religious movement with participating nodes across the world. In this context, the global influences the character of local places, whereas the local explains their shape.

However, the global network of a movement or a group of synagogues is only one of the global circuits that make up the global universe of Judaism. Other synagogues are part of other global circuits. For example, the global circuit of Conservative or Orthodox Judaism, made up of a global block of synagogues, differs from the global circuit of Reform or Progressive Judaism, which have different global architectures of local nodes. As a reflection of this multiplicity, the membership of each global circuit is diversely distributed in terms of geography, social class, and the position of the nodes.

It is one thing to study the form of a global network, but it is another to recognize differences between nodes. The formal global circuit of the group is not identical to its informal or virtual global relations. Therefore, network governance cannot be confined exclusively to the formal realm. These global circuits are different from each other in terms of geography, rituals, and modes of operation, but they are porous because of multiple points of connection of the members who attend services in other sister synagogues, for example. The governance of global Judaism cannot be studied only in terms of its organization, but one must also pay attention to the rapport between agency and doctrine. The porousness of circuits is evident in the extent to which individual members seek guidance in the practice of their faith and the interpretation of doctrinal matters from religious leaders outside their particular circuits. Daniel J. Elazar provides a useful typology of these individuals and institutions that function as bridgeheads or intermediate structures that connect nodes to one another. This includes "the traditional rabbinate . . . the professorial community of Jewish scholars . . . rabbinical seminaries of the Orthodox, Conservative, and Reform camps . . . and the yeshivot and the rebbes of various émigré Hassidic communities."[15] These scholars and institutions serve as hubs that energize and sustain the architecture of the transnational circuits of global Judaism.

THE NETWORK ARCHITECTURE OF GLOBAL JUDAISM

If one speaks of global governance, one must be able to describe the architecture of this network. Architecture is to be understood here in terms of the grammar of social relations: how units are linked to each other and, in the process, how the network governs itself.

As noted above, in Judaism, no single node is in charge of the operation, and leadership is not exercised from the top down. Instead, nodes can individually or as a collective body provide direction to the whole at different times.[16] Nodes may also provide different contributions to the success of the network in terms of finances, agents, ideologies, and operational capability. For network governance to operate, nodes must maintain ongoing relations between themselves through communications that include formal and informal meetings; telecommunications that include telephone, facsimile, and Internet messages; and global agencies—that is, operations that concern themselves with both local and overseas problems and the issues of these transnational communities.

What makes the top-down approach inoperable in relation to Judaism is the absence of one central figure, institution, or bureaucracy solely responsible for the orderly transmission of the faith. Different individuals, such as the rabbi, the scholar, the leader of a secular organization, or even a group of such persons may step in at different times to play a leadership role either directly, by providing religious direction, or indirectly, by protecting or establishing the infrastructure for the practice of the faith. In this sense, one may speak of turbulences in this form of network governance, because perturbations in one part of the network affect the whole and may call for a readjustment of the global leadership structure. That same modulation also emerges at the worldwide level when the intervention of a vibrant overseas community or a new immigration stream reactivates a dormant community.

Network governance takes various forms, a common feature of which is the extension of services from one site to another community in a foreign country. In this context, the network is not hierarchically constructed for the purpose of control, but rather is a series of linkages maintained for the purpose of providing specific services to other communities. As Elazar notes, "the Israeli Chief Rabbinate provides services for a number of small communities (e.g., Iran, India), as does the UK rabbinate of the United Synagogue for some of the communities in countries which once comprised the British Empire."[17] These are communities that do not have their own rabbinate, but seek help from the leadership of a more established religious community or from the homeland.

Although Judaism does not operate worldwide on the basis of a supreme spiritual leader serving as a pope, it does have special units that follow a similar model, such as the rebbes of Orthodox communities, who set the tone for the global networks they command. In addition, network governance is affected not only by religious leaders who occupy religious positions such as that of rabbi, but also by religious leaders who serve in secular leadership positions as heads of associations.[18] The first case is leadership exercised through ritual practices, Talmudic interpretations that guide moral conduct and availability for guidance to the community, whereas the second case is leadership exercised through secular action with a religious connotation and through the linkages that the leaders facilitate because of their relations with other institutions in the fold.

The units in the network are porous in the sense that they influence each other, as when a Sephardic rabbi serves a congregation with both Sephardic and Ashkenazi members or when a specific ritual is borrowed from another Jewish tradition or prayerbook. A singular feature of network governance thus is the osmosis that occurs when a synagogue's ritual practices incorporate a specific religious feature of another community. Allon Gal notes that, "'A Prayer for the State of Israel,' based on a text written by the Chief Rabbinate of Israel, is included on pages 416–417 of the prayerbook [*Siddur Sim Shalom: A Prayerbook for Shabbat, Festivals, and Weekdays*]. Interestingly, this addition to the Conservative prayerbook is mirrored in the Birnbaum Orthodox *siddur* as well."[19] This example shows the porousness of the networks and nodes and how one may influence the other.

THE GLOBAL INFRASTRUCTURE OF JEWISH DIETARY PRACTICES

In a series of interviews with rabbis and ordinary Jews in both Israel and the diaspora (the United States, Canada, Britain, and France), I collected their views on the global governance of *kashrut*, or religious Jewish dietary practices, in order to understand the global dimension of the commercialization and consumption of kosher products.

Kosher products are stamped as such to allow the faithful to distinguish them from other food products. Regulating, producing, commercializing, and consuming kosher products involves a long process that includes the participation and collaboration of several people in different countries at different times. Once certified as kosher, a product remains kosher as long as nobody touches it or opens the container prior to consumption.

The global mobility of kosher products is a new phenomenon. For centuries, the production, commercialization, and consumption of kosher products were local undertakings. Each community produced its own kosher products or purchased them from a contiguous community. Today, because manufacturers supply the high demand by selling their products all over the world, they use a stamp unique to their company so that clients can easily recognize their brands. Stamping also feeds competition between local distributors.

According to an Israeli town rabbi who provides a broad view of kosher in the context of globalization and diasporization, the *kashrut* authorities in Israel and the diaspora are made up of rabbis. Because close to 60 percent of the meat consumed in Israel comes from abroad, a few rabbis travel abroad to administer the *shoreth* (religious/ritual slaughter) and supervise the animal slaughtering process so that it follows proper rituals, all before shipping the meat to Israel.

The commercialization of kosher products accounts for their widespread distribution among Jews and Gentiles. People genuinely believe that these items are better than regular products and contribute to the improvement of one's health. The bulk of the *kashrut* business is in private hands. It is a private business that meets both religious needs and the global market for health foods at large. According to a town rabbi in Israel who once lived in the United States, the "Circle K" brand is the most popular line of kosher products there, and for some, it is more respectable than the "Circle U" brand.

The most popular and available kosher products in Israel—with some exceptions—carry the Orthodox stamp. Side by side with *kashrut* products supervised by the rabbinate are the other products deemed to be kosher that mostly Conservative, Reform, and non-practicing Jews buy. However, the Orthodox establishment and its networks of Israeli practitioners often contest the kosher status of these other commercial products.

A town rabbi who lives in the vicinity of Jerusalem told me that he once visited his mentor, a very old ultra-Orthodox rabbi who lives in a building with apartments also occupied by other ultra-Orthodox rabbis in an Orthodox Jerusalem neighborhood. At the end of the visit, the old rabbi murmured to his guest—my interviewee—that although all of his rabbi neighbors are good friends, they do not eat at each other's tables, because each rabbi belongs to a different sect, and each one doubts the Orthodox quality of the other's *kashrut*.

The global commercialization of kosher products facilitates their communal use and enables broader access and competition in this market arena. This reflects another way in which Judaism in general and synagogues in particular have become globalized. In a supermarket in Israel, one may easily find the following brands of kosher products, among others: "Circle U" (Orthodox Union), "Circle K" (backward C with a K inside), "TK," "Israb" or "Badatz," "Star K" (a star with a K in the middle), "OK" (Lubavitch), and "MK" (Montreal Kosher). Some people buy a product because of the kosher brand, others do so based on the reputation of the rabbi who controls the court or on the recommendation of a local rabbi, and still others may purchase a kosher product because of customary practices or simply because of availability. The law of supply and demand interjects itself in the acquisition of products.

Because of the products' diasporic origins or their own sect's affiliation, congregants often disagree on the type of kosher foods they can serve for communal events at the local synagogue. Those who grew accustomed to a specific brand of kosher food in their country of origin prefer to eat the brand they trust, whereas others may be familiar with other brands. When such a disagreement occurs, the congregants may call upon the rabbi in charge of the synagogue to resolve this local conflict. Global commercialization thus provides ample choices in kosher food types made available to the public and shifts the congregant's posture from using local kosher products supervised by the local rabbi to using international food products for which the identity of the rabbi who supervises the process may not be known.

The kosherization process is not uniform from one country to another or from one religious movement to another. It all depends on the interpretation of the law upheld by a Jewish court, and different types of courts have different standards for what they consider kosher. One or more rabbis in charge of a Jewish court authorize or supervise the making of a kosher product and the decision to ascertain whether it actually is kosher. In the end, purchasing kosher foods is a personal decision, and each consumer personally decides which products he or she deems kosher. Thus, the globalization of kosher derives not from the uniformity of the process throughout the globe, but from the diversity caused by the geographical expansion of the faith.

The politics of kosher labeling and its meaning in the Jewish community have a global dimension in that each label is a global outpost by which a group is recognized and by which the clientele reaffirms its religious identity. However, the application of kosher labels and their multiplicity are recent phenomena. In the recent past, local rabbis decided what products were kosher. In essence, kosher labeling has become a mechanism of identification for groups within Judaism.

The globalization of kosher products has been a slow process. As already noted, prior to the late twentieth century, kosher foods were produced locally for local or national clienteles. They were marketed in stores in specific enclaves where a community of religious Jews resided. Even then, reaching such groups often proved difficult, because not every entrepreneur wanted to sell kosher foods, either because secular Jewish businesses did not want their stores to be identified as Jewish stores or because other businesses calculated that kosher products were not viable in terms of profit maximization. Such products were more suitable for distribution through religious Jewish businesses because they were known as such, were located in the midst of a Jewish enclave, and had a ready-made clientele.

Although in the past kosher consumption in Jewish diasporic enclaves was a local phenomenon, even then there was a slow progression toward globalization as isolated communities depended on foreign synagogues to send specific items (kosher wine and matzos for Passover), foreigners brought items with them during their visits to such enclaves, or purchased products and took them home abroad. This was a sporadic, non-routine form of globalization.

After the fall of the Soviet Union, when additional markets opened up and products and people became more globally mobile, the governance of the production and distribution of kosher foods transitioned into a global phenomenon. The local composition of the rabbinical court, or Beth Din, responsible for converting products into kosher foods and deciding what is or is not kosher, the foreign origins of products, the ordering of kosher meat from specific countries (France, Israel, Argentina, the United States), the global marketing and distribution of products, the global publicity for such products through Jewish Directories or Yellow Pages, the rise of e-commerce sites where such products can be purchased directly online, and the global manufacturing of these products all point to the network governance of the process, a central feature of global Judaism.

Victor B. Geller spells out the globalization of the production of kosher products and explains how the process, with its multiple tentacles, is choreographed.

> A plant might be located in a distant and remote location. The production process might have been very complex. Ingredients were shipped by a half dozen different suppliers, each in a different location. Not only would the primary plant need *hashgachah* [kosher certification] but its network of suppliers would also need supervision. Salaries, travel costs and hotels for a team of *mashgichim* [supervisors] often added up to more than the budget-minded client was prepared or able to spend. The (U) [the Orthodox Union] signed more than a few contracts under which it subsidized the *hashgachah* it offered.[20]

A letter written by Julius Berman in 1996 and reproduced by Saul Bernstein spells out the global scope of the kosher enterprise. In it, he describes the governance of the expansion of *kashrut* and its globalization:

> The expansion of the (U) Kashruth certification service into over 45 countries, with close to 3,000 plants under supervision; the arrangement with the Chief Rabbinate in Israel sanctioning the establishment by our Union of an office in Israel for our (U) supervision and certification, jointly with that of the Chief Rabbinate, of Israeli-manufactured products to be sold throughout the world; the expansion of our internal Rabbinic headquarters officials to 30 Rabbinic personnel.[21]

The publicity for selling such products has a global scope, involving attempts to reach observant Jews, wherever they may be located. It is both a service to the community and a business with global tentacles. To reach the so-called kosher market, "the Kashruth Division published hundreds of thousands of copies of the (U) Kashruth Directory listing all (U) products and facilities as well as hundreds of thousands of copies of the (U) Kosher for Passover Directory. The Kashruth Division of the Union of Orthodox Jewish Congregations of America is the world's hub of Kashruth information."[22]

Because of its size, the American kosher industry, aided by its global publicity and distribution network, has been successful in penetrating Jewish communities outside the United States. The demographic strength of the population, the relative cost of products, and the competitive local market often influence the commercialization of kosher products and their worldwide distribution. The price difference between Israeli and French products has created a niche for the less expensive Israeli items among the French Jewry.

The globalization and global governance of kosher foods has delocalized the production, distribution, and consumption of Jewish dietary products. Here again, network governance provides the technology used to develop production, plan distribution, generate publicity, and enhance consumption. It is evident that, because kosher products, shops, and restaurants were not always available in places with smaller or dispersed Jewish communities, this absence negatively influenced consumption to the extent that large groups of practitioners were not maintaining a kosher lifestyle individually, with their families at home, or collectively.

With the globalization of production and distribution, local communities are no longer dependent on their rabbis for the acquisition of kosher products. Now kosher products are produced or manufactured in large quantities in such cities as Jerusalem, Paris, Berlin, London, New York, Buenos Aires, Montreal, and Toronto, and such items are publicized through Jewish Yellow Pages, regional newsletters, newspapers, and the Internet, as well as distributed throughout the globe. The community does not know the rabbis who make such products kosher, as they did when such items were produced locally. These products are now made available almost everywhere in both Jewish-owned stores and Gentile operations. In Gentile stores, they are regarded as ethnic foods and no longer grapple with the stigma attached to the commercialization of exclusively Jewish products. There is now a large and protected market for such products, and one need only to look for a sign on the product identifying it as kosher. The product may be considered kosher in one locale, but its distribution is boundless.

Globalization has reinforced two peculiarities of kosher products. Because only rabbis can make items kosher, non-Jews cannot easily sell similar products to penetrate the market, which remains, by and large, a protected niche. Globalization has also reinforced the target market, because mostly Jews buy such products. The other peculiarity of kosher food is that it is more expensive than non-kosher foods. There is an extra cost to make it kosher, because kosher shops must pay a fee for a rabbi or Beth Din to certify their operations.

Network governance of the production, sale, and consumption of kosher food is characterized by the authority of a group of rabbis who have control over the process. Such individuals ensure that dietary codes are followed and that proper rituals are used. The distribution would not succeed worldwide without storeowners who are willing to participate in the network of distributors and collaborate with the headquarters of the operation. The manufacturing of these products can also be global, because some people may rely on products from other countries.

The global production, distribution, and management of kosher products are not centralized processes, but proceed from the authorities and size of Jewish communities in each country and each place. This assessment reflects the condition of the bulk of products, because different communities adhere to different national dietary practices, cultural customs, and ritual traditions, which affect mass-market, commercial, and globalization factors.

There are several factors that account for the globalization of kosher products when authorities in one country take charge of communities in another. Of course, they are likely to favor the kosher products with which they are most familiar. The religious ties between these communities favor the penetration of the local market with products that have the imprimatur of religious authorities elsewhere. Headquarters or subsidiaries in several countries are used to manufacture, control, and distribute these products.

However, when products made in one country are distributed and commercialized elsewhere, these transactions are entrepreneurial activities over which the religious authorities have no control. They follow the laws of market of supply and demand and robust competition. As a result of immigration, in Israel we now find kosher products available in supermarkets from various other countries, including the United States, France, Italy, and Canada. People who previously lived in one of these countries or who belong to a specific religious movement often buy these kosher products.

More recently we find kosher products sold online, in which cases their places of origins may not be important, as long as they are kosher. One must distinguish the Kosher Yellow Pages online (www.theonlykosherdirectory.com) from kosher groceries where one can order online (www.kosher.com).

From a managerial standpoint, the production and distribution processes of kosher products follow several paths. Some are managed at the national, regional, state, or city level. At the city level, one can think of the kosher process controlled by the Jewish Community Council of Montreal (MK), the Star-K Kosher Certification of Baltimore (Star-K), or the Ufficio Rabbinico Di Trieste (Trieste, Italy). At the regional level, one may mention Vaad Hakashruth of Northern California; at the state level, the Orthodox Rabbinical Council of British Columbia (Canada), the Chief Rabbinate of Quebec-Kashrut Commission; and at the national level, Sistema Kosher (Mexico), the Kosher Supervision Unit—Quality Control Department, Buenos Aires (Argentina), the Kashrut Department, Beth Din of Johannesburg, the Union of Orthodox Synagogues of South Africa (South Africa), and She'erit Yisrael Kosher Certification (Bnei Brak, Israel). Some have their headquarters in one country and distribute their products worldwide, as in the case of the Union of Orthodox Jewish Congregations (U) or the Organized Kashrus Laboratories (OK), both located in New York City; some operate as a two-country operation as in the case of the Central Rabbinical Congress of the USA and Canada or the Services International Kosher Supervision Ltd. (Jerusalem, Israel, and Torino, Italy); and sometimes the operation has headquarters in more than one country, as does Bdatz Mehadrin, which is located in Rechovot (Israel), Antwerp (Belgium), and Lakewood, New Jersey (United States). In countries with chief rabbis, such as the Office of Chief Rabbi of Bnei Brak (Bnei Brak, Israel) or the Kashrut Division of the London Beth Din, Court of the Chief Rabbi (London, UK), this higher authority exerts control. These various forms tend to reflect the history of the Jewish community in each country and the level of its religious groups' fragmentation.

GLOBAL TRAINING FOR GLOBAL LEADERSHIP

The functioning of the global network can also be seen in the field of religious education. Like Rome and Mecca, Israel serves as a pole of attraction for the diaspora in search of its religious roots and as a place for training Jews to minister for overseas congregations. Elazar notes that "the Indian Jewish community, which has no rabbi, periodically sends young men to study in Israel to learn to be at least *shohetim* [kosher slaughterers] and *mohelim* [circumcisers]."[23] Young students from around the world travel to Jerusalem for further religious training. This ensures proper training for the adherence of future religious leaders to the mainstream interpretation of the faith, access to religious education by the local communities around the world, and the establishment of transnational linkages that support the infrastructure of local synagogues. In his survey of the organizations of Jewish communities, Elazar reports that "a very high percentage of indigenous Latin American Jewish teachers have received all or part of their training from the Hayim Greenberg and Gold Institutes in Jerusalem, the first for secular and the second for religious schools."[24]

American students, too, spend time in Israel to complete their training in theological studies, irrespective of their Conservative, Orthodox, or Reform affiliations. For example, Rabbi Israel Goldstein notes that "Conservative Judaism is represented in Israel through Neve Schechter in Jerusalem, where students of the Jewish Theological Seminary of America spend one year in study, directed by Rabbi Reuven Hammer."[25]All branches of Judaism are beginning to follow the practice of sending theological students to study in Israel prior to their rabbinical ordination. Adam S. Ferziger notes that "in practice, while RIETS [the Rabbi Isaac Elchanan Theological Seminary, New York] students are *encouraged* to spend a year of study at Yeshiva University's Gruss Institute in Jerusalem, YCT [Yeshivat Chovevei Torah, New York] candidates *must* spend one of their four years in Israel."[26]

Some schools in Israel prepare their graduates to serve as rabbis abroad. Ferziger remarks that "JSRS [Joseph Strauss Rabbinical Seminary, Israel] attracts students from all over the world who participate in a four-year course of study that culminates in rabbinic ordination. The graduates are then expected to serve as rabbis in Diaspora communities, including in the United States." In other, similar cases, a Hassidim community may use its yeshiva's rabbinical ordination/leadership program "as the most effective way to supply manpower for its centers in the diaspora."[27] Some students go to Israel to immerse themselves in their religious heritage and, upon returning to their homeland, become part of the team engaged in revitalizing their community.

Several institutions are involved in this crisscrossing movement, which sends people abroad to do missionary work or to facilitate global mobility for study purposes. For example, the Department for Torah Education and Culture in the Diaspora

> maintains educator-emissaries in the Diaspora, who constitute its principal line of contact with Jewish communities and their educational institutions. In 1981–82, 221 such *shlihim* [emissaries] were abroad. The largest numbers (56) were active in Canada, followed by Australia (31), Brazil (23), the United States (22) and Great Britain (20). About 70% were maintained by the host community. Of the balance, expenses were divided between the Department and the community in most cases; only in the remaining 7% was the Department solely responsible for the cost.[28]

Moreover, "the Department also enables young men from abroad to enroll in yeshivoth and rabbinical seminaries, and young women in Torah institutes. Under this program, which receives support from the Ministry of Absorption's Student Authority, some 2,300 young people study each year in Israel."[29] In addition, Elazar reports that, "the Israeli presence in Latin America is especially vital in the educational system. Between the two education departments of the World Zionist Organization, approximately 150 Israeli teachers are regularly present in the region, fully or partially paid from Israel or world Jewish sources."[30]

A former head of the Jewish Agency Education Department explains some aspects of the mobility of rabbinical and non-rabbinical Jewish students and how they are being trained for global leadership. Every year, when Jewish students travel to spend a semester or a year of study in Israel, their parents or scholarship funds pay for the tuition. Those who enroll at Yeshiva University pay tuition to the university, which shares the money with the Israeli program.

While some of these students undergo training to become rabbis, others study just for their personal spiritual growth so that they can return home with skills to pursue any profession they want. Basically, each sect has its own school to which they send people in Jerusalem. For example, students from Hasidic families travel to Israel to study in a yeshiva operated by their sect. Usually, one pool of schools in the home country sends their youngsters to schools for their specific counterparts in Israel. Orthodox students go to a yeshiva to study for a year and Conservative movement students go to the Schechter Institute of Jewish Studies.

A Jewish student from Los Angeles, who would later serve as a rabbi, and a foreign-exchange student from France, both attending rabbinical school in Israel, provided more details. The female student, who would soon become a rabbi and intended to return to the Los Angeles area, said that students from the University of Judaism (UJ) in Los Angeles and the Jewish Theological Seminary (JTS) in New York attend the Israel-based program of the Schechter Institute. They travel to spend a year in Jerusalem. While the majority of students at Schechter are Israelis, the institute also recruits students from the United States, France, and Argentina. Like the Americans, these other students return home after their internships in Israel.

The Conservative Cantorial School in New York has its students study abroad for at least a semester and, on occasion, for one year. Not all of the students spend their year of biblical study in Israel; some go to schools in England, or even in the United States. The young men are not the only ones who study in Israel for their own spiritual growth, to become a rabbi, or to attend a secular career path at a university. There is an institute in Israel that is open exclusively to young women. For example, young French women spend a year, and sometimes more, studying in Jerusalem and later return to France. They arrive for special training so that they can become directors of Jewish schools in France. They will not be rabbis, but they will direct schools with a Jewish religious orientation.

THE GLOBAL MOBILITY OF RELIGIOUS LEADERS

As happens in Catholicism and Islam, religious Jewish teachers travel to other countries to help schools and congregations. There are several reasons that explain why religious leaders may move from one country to another to pursue their ministry. A town rabbi in Israel explained to me that the Jewish Agency and the Rothschild Program send rabbis to help struggling communities in Eastern Europe—for example, after the collapse of the Soviet Union. Missionaries also are sent abroad to proselytize non-observant Jews. For example, Geza Komoroczy observes that "Jewish proselytizing is taking place in Hungary. . . . The new proselytizing has as its pool those former, now secularized Jews, emotional or cultural Jews, and half-and-quarter Jews who now may come nearer to Judaism and the Jewish community."[31] Some are hired by local congregations and emigrate to serve their new flocks; some travel to Israel from abroad to study the Torah and the language; and others migrate to other countries to improve their economic conditions. Unlike town rabbis, who earn enough to make a living, the other rabbis receive modest wages and often must emigrate to find positions abroad. Other rabbis are mobile because they are selected or appointed by foreign communities.

Rabbi migration routes may change over the years, but in any specific period, there are more rabbis migrating to one specific region or country than to others. A Canadian Israeli who has a medical practice in Jerusalem offers the following observation: "Several friends told me that it is rare that an Israeli rabbi is appointed to a synagogue in North America. First of all, there are a lot of rabbis in North America now, but also most rabbis who live and graduate in Israel are Orthodox. By and large, the Orthodox in North America prefer to hire a North American rabbi who speaks English fluently and who understands their way of thinking, because most North American Orthodox are much more liberal than the Orthodox Jews in Israel."

Thus, speaking the local language is an important qualification for migration. In a conversation with a French Jewish teacher who translated a prayerbook from Hebrew to French, he stated unequivocally that certain French rabbis have served in Switzerland, Belgium, and even in the island of Reunion. Their mobility operates within the Francophone world.

However, any node in the network may establish its own regional or global field of influence and may enter the global network as a bloc or as a separate unit with its distinct agenda. Missionization, when a rabbi leaves to provide religious leadership to an overseas community, forms these regional or global subsets. The strength of the relations between the New York Jewish community and the Argentinean Jewry must be seen in this light. As Elazar notes, "the Conservative movement sent a rabbi to Argentina who, after considerable struggle, managed to build a substantial network of congregations affiliated with the United Synagogue, with ancillary activities including a theological seminary, a Ramah camp, and a variety of educational program."[32]

The American Jewish community has played a major role in reaching out to overseas communities by providing religious assistance to postwar Jewish communities in Europe. This was done by opening their seminaries to Jewish students from other parts of the world and by sending rabbis to Canada and Latin America. These processes have enabled the community to develop a global field of influence. Elazar notes the American provision of

> rabbis and teachers for those communities unable to train their own. The American Jewish community began by taking a few students from Latin America, Europe, and even India into its seminaries in the immediate postwar period. Today American rabbis are themselves going out to serve communities overseas. Mention has been made of Argentina. The chief rabbi of Sweden is a Philadelphian who is a graduate of the Jewish Theological Seminary. Japan invariably has an American rabbi.[33]

Since the end of World War II, Europe has also become a theater of action for the American Jewish diaspora as it provides religious leaders to reinvigorate these remnant communities. Thus, "the recent rabbis in Gothenburg and Stockholm and the cantor and baal korah of Stockholm are all from the USA."[34] American Jewish influence is not confined to any specific country, but is widespread throughout the European Union. David Clayman notes that "American and Israeli rabbis and educators are working in communities throughout Europe. . . . American-trained rabbis today are serving communities in Sweden, France, Poland, and the former Soviet Union. In most instances, these rabbis are Conservative and Reform. . . . On the other hand, Israeli spiritual assistance to European Jewish communities is largely from the Orthodox establishment."[35]

American Jewish influence on Canadian Jewry is also palpable because established links have existed for many years between these communities, and Canadian Jews rely on American Jewish theological centers for training and on American-trained rabbis. As Sol Encel and Leslie Stein remark,

> Judaism in Canada is a branch plant operation. Reform, Conservative, and Reconstructionist Judaism are all completely dependent on their American counterparts for infrastructural support, and, more important, for the major rabbinic seminaries. These rabbis in Canada must be trained in the United States. There are some ultra-Orthodox rabbinic seminaries in Canada, but the larger, modern Orthodox institutions such as Yeshiva University are likewise south of the border. The majority of pulpit rabbis in Canada have historically been American. Only recently has there been an increase in Canadians who receive ordination in the United States and then return to Canada.[36]

Even more established Jewish communities, such as those of England, have also felt the power of American influence. For example, Marlena Schmool finds that "Jewish booksellers report that approximately 95 percent of the Judaica (in its widest sense) and Jewish books sold in Britain originate in the USA."[37]

The mobility of religious leaders tends to flow toward the sites where the movement is already established, unless missionaries are sent to a different place to proselytize and reactivate the faith of brethren. Daniel J. Elazar and Rela Geffen, in their study of the Conservative movement, note that:

> When Rabbi Meyer moved to New York, he brought graduates of the Seminario Rabinico to work at Congregation Bnai Jeshurun in Manhattan and created greater interaction between the Seminario in Buenos Aires and the JTS in New York. JTS faculty teach there regularly, as they now do in Budapest as well. . . . Several of the Masorti rabbis in England were ordained at Machon Schechter in Jerusalem. Rabbis from Machon Schechter also are serving in the former Soviet Union and in reconstituted communities elsewhere in Europe.[38]

However, some groups sent missionaries abroad to enlarge their membership and to establish new centers of the faithful. Encel and Stein note that "missionaries of Agudat Israel, Lubavitcher, and Satmar Hasidim were dispatched [by Orthodox rabbis in the United States] to the principal cities of

Latin America, founding *yeshivot* (religious schools) to attract the sons of the unaffiliated. The Lubavitcher Hasidim (known also as Chabad) have 13 missionaries (to Jews only) in Argentina today, not only in Buenos Aires, but in regional centers as well."[39]

GLOBAL JUDAISM AND THE NATION-STATE

Although network governance in Judaism is highly influenced by the behavior of local nodes, the transnational nature of the network, with students from around the world receiving training in Israel and rabbis dispersing throughout the nodes of the network and creating new ones, necessarily involves relations with the public policies of local nation-states. The local units that constitute the nodes of the global network not only are distinct and interlinked, but they also have different modes of aggregation at the national level and different forms of participation in and contribution to the network on that basis.

In particular, any state that provides social and territorial infrastructure to nodes in the network may enhance or impede the manner in which transnational linkages are undertaken or prevent their full deployment. In this light, one might say that network governance, because of these state constraints, is limited in the manner in which it can operate. The Consistoire, established during the Napoleonic era on the basis of a vertical axis that governs the relations between synagogues and the chief rabbi of France, is an excellent example of how the local nation-state can help structure and therefore limit the parameters of action of a local synagogue. This structure, very much inspired by the hierarchical structure of the French Catholic Church, nevertheless recognizes the need for an autonomous religious space for Jews.

Synagogues in Britain are grouped into a collective religious body called the United Synagogue and function under the leadership of the chief rabbi, who supervises the appointment of rabbis to local congregations. Board members of the United Synagogue are part of the power-sharing mechanism in the governance of this national structure. In the United States, the power of governance shifts to the local branches, regardless of Orthodox, Conservative, Reform, or Reconstructionist orientation. As Elazar puts it, "every congregation, no matter how committed it may be to its movement, hires its own rabbinical staff, under its own terms, in what amounts to a free market situation."[40]

With regard to their relations with the nation-state, the nodes of Judaism's global network can be distributed into five categories: established communities, new communities, remnant communities, institutionally incomplete communities, and the State of Israel itself as it functions as a node in the network. The established communities have a long history in the country where they are found. Because of anti-Semitism, some communities began as enclaves or ghettos, although some members dispersed and socially integrated within the larger society after emancipation. The new communities emerged after World War II as a result of persecution, immigration, or homeland independence in the cases of Morocco, Tunisia, and Algeria. A remnant community is what is left after a large number of people have died or emigrated to a safer country. Institutionally speaking, incomplete communities do not have all the institutions that can respond to the needs of the community. For example, they lack synagogues or rabbis to staff the synagogues, lack Jewish schools to teach the history, culture, language, and the basics of the faith, or lack social agencies such as hospitals, welfare bureaus to care for the poor, the elderly, and those in need of help. Stockholm is a good example of such a case. It is "representative of small communities which import Jewish scholarship and knowledge rather than create them for themselves and other communities."[41] In terms of their histories and institutional strength, these communities have relatively more or less weight within the global network, more or less power within the social and political structures of the nation-state, and require more or less assistance from other nodes in the network.

Finally, the State of Israel, itself a member of the community of nations, plays a unique role as a node in the global network of Judaism. In Judaism, global governance is centralized when it pertains to the definition of Jewish identity, and this node thereby polices the borders of religious Jewishness. The Israeli rabbinate and the Knesset that it influences on religious matters play central roles in deciding who is or is not a Jew. This role affects not only Israelis, but also Jews in the diaspora and those who have converted to Judaism. This is seen most clearly in the case of those who travel to Israel and must interact with Jewish institutions. One can think here of the application of the law of return. As Elazar says, "the only issue in which all Jews may be subject to some kind of binding decision making is in determining Jewishness itself, that is to say, 'who is a Jew,' where, because of the influence of Israel, the decisions of its authoritative institutions on this question are authoritative for the Jewish world as a whole."[42] Other sectors feel and submit to the authoritative voice of Israel as well. Clayman notes that, "what has happened in many communities throughout Europe is that the official Jewish community looks to Israel and to the Rabbinate in Israel for direction, supervision, and services."[43]

THE GLOBAL GOVERNANCE OF JEWISH SECURITY

The global governance of the faith includes the governance of the security apparatus employed to protect people, buildings, and sites to prevent vandalism and the desecration of cemeteries and synagogues. Although the State of Israel and various nation-states contribute to the creation and functioning of this apparatus, it is largely a local matter. In Europe, in general, the public powers assure the security of synagogues and sites, as well as schools. For example, in Paris, the French police assure the security of synagogues and schools. Despite this municipal police oversight, communities have developed their own volunteer groups and contracted paid agents to provide security on the shabbath and the high holy days.

Informant interviews gave a sense of the magnitude of the problem and provided individual interpretations of the issue of governance. A Canadian Jew from Toronto stated that synagogues there have security guards, not members of the municipal police force, who are paid for by the synagogues. According to this informant, there is a high level of cooperation between Israel and Interpol, and if the government of Israel discovers an impending threat against a specific synagogue or community, it may contact the government of Canada. He went on to say that in Canada, the local police cooperate with the synagogues, and the surveillance of the synagogues is done mostly when they are in use. If surveillance were to be done around the clock, the cost would be exorbitant. Instead, synagogues hire guards to do a rotation of duty only when the building is open for religious services or meetings.

An American rabbi who spent a year studying in Israel commented on the evolution of the security system that protects Jewish people and property in Boston. He stated that if the FBI, CIA, or Homeland Security believe that synagogues need to be on higher guard, they disperse this information to people who are particularly targeted by telling key organizations, such as the Anti-Defamation League and the Jewish Community Relations Council, who then relay the information directly to the synagogues.

In Boston, some synagogues have hired security since 9/11. Because security has become such a priority, synagogues have cut the budget for other things in order to pay the guards. Since 9/11, Jewish institutions in the Boston area have started to conduct themselves like places in Israel: checking people, making sure that things are under control, and making sure that there is only one entrance instead of numerous ways to enter a building. For Yom Kippur, some synagogues hire a security team. Some hire a security system on a permanent basis, whereas others hire on a need basis. Another American rabbi informed me that in Los Angeles, synagogues routinely hire a security guard to protect the congregation. Just as the synagogue's budget includes general maintenance, it also includes security costs.

Because the security system is handled differently from country to country, I interviewed a French Jew from Marseilles about the protection of synagogues there. In Marseilles, the security apparatus is a private organization that works parallel to the municipal police. This arrangement functions only on high holy days such as Yom Kippur to secure the synagogue and other places of assembly. This structure has resulted from the Jewish community's commitment to assure the security of Jews on the site, but because there has been no money to pay those who provide security, the private organization has been composed entirely of volunteers. Even before

9/11, there were always police cars at the entrance of the synagogues and in front of other Jewish institutions during the high holy days. In previous years the security was more discreet, but there was always a security presence at the synagogues. There was usually a body check, but much more modest than the elaborate search in collaboration with the police that is seen today.

Another French Jew, a retired teacher, shared with me his observations of the functioning of the security apparatus in Paris and Istanbul. Whether in Istanbul, Paris, Berlin, Kiev, or London, one cannot easily enter a synagogue without showing a piece of identification and undergoing some body check. Some synagogues have their own security camera to monitor suspicious behavior. This check is done whether one is a Jew or a Gentile.

Although the creation and functioning of these security apparatuses is largely in local hands, since the Six-Day War in 1967, maintaining an equilibrium in the network governance of the security apparatuses has depended on the crisis level in Israel, which may demand the attention of the diaspora. A crisis in Israel directly affects the financial stability of any synagogue because it can change "the direction of local fund-raising efforts in every diaspora community."[44]

THE DIGITAL GOVERNANCE OF GLOBAL JUDAISM

Information and communication technology has significantly contributed to the global and network governance of Judaism.[45] Although there is no central mechanism or authority that influences the shape of digital Judaism, as in the case of global Catholicism, there are particular global networks that attend to the interests of its specific branches. Each global movement, including Orthodox, Conservative, Reform, and Humanistic Judaism, uses IT to communicate, share information, provide services, and to a certain extent guide the path of the affiliated synagogues. Only one ultra-Orthodox community discourages its members from using the computer to avoid being exposed to unorthodox practices.

Because each movement uses IT to meet its own particular objectives, each global platform differs from the others. In each movement, however, a center gives coherence to the whole in an effort to maintain the unity of the group, despite its geographical diversity. The center shares information, provides guidance and services, and remains in contact with associates. This can be seen, for example, in the case of the Conservative movement.

The United Synagogue of Conservative Judaism has established its own Department of Information Technology, providing digital leadership in three prominent areas. First, organizational information is made available through its central server to all the offices in the network. These include the headquarters of the movement, its regional operations, and the local schools and synagogues that fall under its umbrella. Exchange of information between the headquarters and the satellites is carried out through this digital infrastructure, and local offices download files they need from the central server. Second, information garnered by surveying the needs of all the synagogues in the network is now part of a database that is accessed by the headquarters to identify and respond to the needs and aspirations of the membership of the Conservative movement. And third, the central office, through its IT Department, provides guidance to local congregations, helps evaluate their computer needs, and serves as a facilitator in their efforts to select and acquire both software and hardware. This strategy helps save money locally, makes the network interoperable, and creates resources that can be shared among the constituencies.

One aspect of the deployment of network governance is the way in which the central office helps manage the assessment, acquisition, and use of computers in all the nodes of the global network. This reinforces the legitimacy and authority of the center, harmonizes computer practices throughout the network, and, in the process, links every node to the central office. Through this arrangement, as information flows from node to node, the network maintains its distinct identity different from those of others.

There are two forms of leadership in digital network governance: active and reactive. Leadership is active when a central office prepares content and distributes it to the nodes. This happened, for example, when the Orthodox Union, through its Department of Community Services, developed an Internet safety CD for parents affiliated with its network of synagogues. The CD provides practical advice to parents so that they can be better prepared to monitor the Internet use of their children. Leadership is reactive when individual practitioners phone or e-mail questions to the center, asking for guidance or services. According to Menachem Friedman,[46] Hasidim living in various cities in the Western hemisphere have requested such advice through electronic means from the headquarters of operation of the movement.

Like global Catholicism, Judaism uses the Internet for outreach and to proselytize mostly among Jews who have left the faith. For the success of this strategy, the Lubavitch Movement, for example, has developed "a worldwide network of virtual Jewish centers at Chabadonline.com." The Lubavitch Movement, which has a global structure of network governance, also uses IT to connect centers to each other and to the headquarters. They make texts available in different languages to members through their websites, and they provide information on kosher dietary practices and restrictions as well as services to the global community through, for example, the "Ask a Rabbi" feature, which allows people to get answers to their religious questions.

The digital governance of the Lubavitch network is done through a distributed management system whereby webmasters of sites located in different countries coproduce the output they make available to the global faithful. The global organization of the network governance system among the Lubavitchers was spelled out by Mike Kamber when he noted that "the Lubavitch team includes an editor in Israel and a programmer in the Ukraine, but the content decisions are made at Lubavitch World Headquarters in Brooklyn. Once web content is created by the team, they upload the material onto servers housed in New Jersey, and mail a list of available pages to Lubavitch communities on every continent."[47] This form of digital network governance reflects how input from various sites contributes to the outcome of the process and also how resources or talent from various congregations are put to work for the benefit of the network.

The Lubavitch Movement's style of digital network governance is exemplary in the way it uses various sites in the production of content. It may end up serving as a model of decentralization and effective use of IT for other branches of Judaism and other global religions to emulate.

ISRAEL AND THE UNITED STATES AS PIVOTAL NODES

Since the Holocaust, Israel and the United States have emerged as pivotal nodes in the global network of Judaism because of their status as the two of the largest sites of Jewish settlement, what Jerusalem represents for Jews, and the extensive wealth among U.S. Jews.[48] These pivotal nodes exert their influence on the larger community in complementary, sometimes similar, but not identical ways. For example, both develop subset areas of influence where other communities seek their leadership. The U.S. Jewish community has provided rabbis to Canada, Latin America, and the European Jewry.

The U.S. Jewish diaspora constitutes a pivotal node in the global governance of Judaism because it provides support to Israel and overseas Jewish communities, because Jewish religious movements have influential centers in the United States, and because ordained rabbis train in the United States and now serve congregations throughout the world. Through their international outreach programs, Jewish institutions of higher learning in the United States have also contributed to the strength of the United States as a pivotal Jewish node. One can think, for example, of the Jewish Theological Seminary (Conservative Judaism), which has been involved in the training of a "new Russian Jewish intelligentsia through Project Judaica, its Jewish studies program in Moscow," and its ongoing support and collaboration with the Seminario Rabinico Latinoamericano in Buenos Aires, Argentina.

Israel has been grooming itself to regain its once-prominent place as the leader of the global governance of Judaism. This is not simply because of its symbolism as the geographical center of the faith, but also because it has imposed itself as the ultimate recourse for religious Jewish affairs. It is a projected outcome that has not yet been totally realized. This vision can be detected in the role of the Religious Ministry of Israel, which sees one of its primary functions as "fostering religious ties with diaspora Jewry." This leadership role of Israel is also seen in the role played by the Chief

Rabbinate. It sees one of its responsibilities as responding to "halakhic questions submitted by Jewish public bodies in the diaspora." The global leadership of the Chief Rabbinate is even more marked because it decides on the validity of conversions done in the diaspora. For example, a converted Jew was not allowed to have his marriage ceremony presided over by an Orthodox rabbi in an Orthodox synagogue in Israel because the Chief Rabbinate decided against the validity of the conversion. In so doing, the Chief Rabbinate indirectly has some jurisdiction over diaspora Jews, either by its responses to the halakhic questions they ask or by the religious services it provides when they visit Israel for wedding, conversion, burial, divorce, or counseling. The Chief Rabbinate also decides the legitimacy of a rabbi who performs a religious service, such as a conversion, depending on whether or not he or she is recognized in Israel. No other rabbinate has this kind of broad jurisdiction over the network. As Rabbi Michael Boydon, chairman of the Israel Council of Progressive Rabbis, pointedly remarks, "even when someone has undergone an Orthodox conversion outside Israel, there is no guarantee that they will be recognized as Jewish by the rabbinate."[49]

Israel has become the site where various Jewish organizations have established their headquarters of operation. Some have simply moved their headquarters to Israel, as when the World Union for Progressive Judaism relocated to Jerusalem in 1973. This global institution continues to maintain and interact with its regional offices in New York, London, and Moscow.

As previously noted, some Jewish theological schools have programmed the "Year in Israel" or the "Israel Experience" as a component of their training toward rabbinical ordination. The Ziegler School of Rabbinic Studies (Conservative Judaism) (www.ajula.edu) flatly declares on its website that "to serve as a leader in the Jewish world today requires an intimate familiarity with the state of Israel, its politics, its people, and the challenges it faces."

Israel maintains a pivotal role due to the number of overseas students it attracts every year, the prominent role of the Chief Rabbinate in some aspects of diaspora life, the human and spiritual resources it is able to provide to the diaspora, the opportunities for officials of various movements and associations to meet and discuss things of common concern because their organizations are located in contiguous space in Jerusalem, and the circulation of overseas Jewish students who are able to meet with other students before they return to their respective homeland.

CONCLUSION

Although I am aware of the globality of particular movements inside Judaism such as, for example, the Lubavitchers, I have emphasized the globality of the network.[50] Particularistic globalization gains a new meaning in this context. The network governance of global Judaism sometimes forces the faith to redefine its boundaries or to collaborate for the good of the brethren. Sometimes these relations are formal and entertained by the leadership, and at other times they are entertained by the rank and file in terms of acquiring kosher products from the other sites, or even attending services that are not available within one's own movement because of one's place of residence. In any case, as we have seen, the network governance of global Judaism operates on many levels and, in many ways, even without the agency of a central governing institution.

The analysis of network governance of global Judaism expounded in this chapter, although not exhaustive, nevertheless sheds sufficient light on certain mechanisms of its deployment. Network governance reflects the ways in which nodes are linked to each other for the purpose of consolidating the practices of the faith, enhancing the capabilities of each entity, and ensuring the reproduction of the religion despite the consequences of globalization, fragmentation, and diversity. Judaism as a monotheistic religion has been impacted by the global geographical expansion of its faith-based communities, the differentiation of its organizational and institutional components, the multiplex diversity of its infrastructure, the global mobility of people, goods, and knowledge that it nurtures, the global contribution of learned brethren and Jewish schools of thought to the exegesis of its religious doctrine, and the global networks of relationships that sustain its social reproduction.

Network governance is part and parcel of the organizational make-up of global Judaism. The faith influences in many ways the theological and managerial contours of the network, whereas the network provides an infrastructure for the deployment of the religion and the shape it may take in diverse environments. The governance of such a global network is not carried out exclusively by any single individual, group, site, or bureaucracy, but is multiheaded in the sense that different entities (organizations, institutions, nodes, and individuals) contribute to the process.

Unlike the situation in Islam and Catholicism, network governance of global Judaism has intrinsically self-imposed social boundaries because it concerns mostly the Jewish cosmonation. In this context, ethnicity and religion are embedded in the form and content that network governance of global Judaism takes at any given period. Here again secular and religious networks intermingle at times to provide a backbone to the expression of global Judaism at both the local and global levels.

This analysis of network governance has allowed us to better understand key aspects of the phenomenon by unveiling mechanisms of operation in three areas: the diversity of behaviors of nodes, relations between nodes, and network governance as a procedure of practice. One learns that nodes have their unique histories in terms of formation, growth, maturity, and sometimes disappearance, and the level of their contribution to the network varies according to the position they hold at a specific time. One also learns that networks have their own specific histories as well and that they are prone to friction, tension, competition, cooperation, and extinction. Thus, network governance of Judaism implies the governance of multiple networks, the relations between them, and the collaboration between secular and religious Jewish institutions. Network governance ultimately is made possible because of the religious vision the faith expounds, the global infrastructure of practice it establishes, and the urge to collaborate in order to maintain and strengthen the viability of the cosmonation.

In conclusion, it must be said at the outset that network governance has its own tempo and exhibits a variable geometry of leadership depending on the mode of operation that molds its practice. For example, network governance is activated occasionally for specific purposes such as support of Judaism in a time of crisis, the protection of a diasporic site, or the enactment of a public policy that may negatively impact the practice of the faith and the well-being of the Jewish community. Different combinations of groups at different times come forward to address the problems of the faith or the cosmonation and provide disciplined leadership to the network to ensure its longevity and social reproduction.

3 Global Catholicism and Corporate Network Governance

> The earliest Church was in Jerusalem. . . . Its first leader, according to Christian tradition, was James the Less, remembered as the first Bishop of Jerusalem. On the presumed site of his martyrdom stands the Armenian Cathedral of St. James. In the year 49 or 50, as recorded in Acts [of the Apostles], a local council, meeting in Jerusalem, made fateful decisions for the young church. It was decided that Gentiles could be admitted to the Christian fellowship and that upon conversion they would not be required to uphold the precepts of Judaism.
>
> —S. P. Colbi, "The Christian Establishment in Jerusalem"

Since its formal institutionalization in the West, Catholic Christianity has been governed by a corporate headquarters in communion with bishops, archbishops, cardinals, and patriarchs that constitute its global leadership network.[1] The corporate nature of the network governance of the Church is seen in the convening of various synod and council gatherings to establish the contours of the faith, identify and banish heretical beliefs, and develop devotional and liturgical celebrations. The corporate headquarters has been established and seated in Rome for the past two thousand years, except the short interim period between 1309 and 1377, when Avignon, France, claimed and usurped that prerogative. Rome—first in the Lateran Palace and later the Vatican—has been the official administrative seat of government charged with the mission of providing global leadership to the Catholic faithful.[2] Since the Lateran Treaty of 1929, the very locale where the headquarters is now established, officially known as Vatican City (including the site of St. Peter's tomb, inside the basilica—in the vicinity of which the papal residence is located) has gained legal recognition as an autonomous nation-state with its own legal instruments, foreign policy, security apparatus, and domestic agenda.[3] It is the workplace of the administrators and employees of the Holy See, and large numbers of pilgrims and tourists visit the area on a daily basis.[4]

The independent status of the Vatican as a city-state sets the Catholic Church apart from the practice and experience of other world religions, which cohabit with the state and fall under state laws.[5] In contrast, the Vatican maintains state-to-state relations with Italy and other countries, appoints observers to the U.N. headquarters in New York, sends official delegates to international forums, and has diplomats known as nuncios (with ambassadorial rank) throughout the globe with the mission of advancing the Holy See's agenda and the Vatican state's foreign-policy goals.[6] With the pope as head of state and the spiritual leader of the universal church, the Vatican is also a city that has employees and that must provide city services to its residents, citizens, pilgrims, and tourists. In all of these aspects, it has developed a network governance structure with a corporate headquarters, including a large bureaucracy that has its own unique characteristics, unlike the network administrative leadership of either Judaism or Islam.

The Vatican also differs from the two other monotheistic religions in several ways, because, as the site of the headquarters of the Catholic Church, it is not the place where the founder of the religion was born, resided, died, or was buried. Jerusalem may claim to be the geographical site where the homeland religion was first established, and Mecca may do the same in the case of Islam, however the Vatican City cannot make such a claim, because Western Catholicism is basically a religion governed from a diasporic site. In this specific instance, the diasporic administration has relegated the homeland church (the primitive Church of Jerusalem) to a subsidiary position. The theological and sociological explanations and justifications provided to dissidents are that Jesus selected the figure of Peter to lead the nascent Church, that Jewish dispersion after the destruction of the Second Temple of Jerusalem accounts for this geographical displacement of the headquarters from the homeland to a diasporic site, and that Peter's martyrdom took place in Rome where subsequently his body was buried on Vatican Hill.

As the site of the headquarters of the Catholic Church, the Vatican displays its own peculiarity in network governance due to a corporate structure that mirrors its relations with overseas branches. This chapter shows how the Church's corporate network governance operates on a global scale. It also examines the role of the Holy See in the deployment and functioning of this global managerial process, as well as its effect on the shaping and structuring of Church headquarters.

The sociological literature distinguishes networks from hierarchies as two distinct forms of organization.[7] Here, we will see that the two are not mutually exclusive and we will examine the effects of hierarchy on the structure of the corporate network of nodes.

CORPORATE NETWORK GOVERNANCE

In addition to implying a certain notion of spatial centralization in the organization of administrative work, the concept of corporate governance, as opposed to less centralized forms of governance, implies a hierarchy in the vertical distribution of managerial leadership, a headquarters in reference to satellite entities spatially distributed in other locations, a top-down decision-making mechanism in lieu of a decentralized process, and a bureaucracy that constitutes the backbone of the social institutional infrastructure in contrast to the one-person management model of small family firms.

In the recent past, one seldom thought of corporate governance as a function of a global network of sites, because each firm was located in a specific area. There was a common understanding that network and hierarchy do not mix because of the different logics that characterize these forms or models of business organization. In recent years, however, as reflected in the business literature, the rise of multinational corporations has led to a shift in the way in which we study transnational organizations.[8] As established networks of sites, it became obvious that such dispersed and interlinked entities must be managed or administered through network governance. However, it was not until the information technology revolution, with its panoply of computers, cell phones, and other mobile digital devices—all of which have Internet access—that communications among units distributed in various locations became both instantaneous and interactive. Information technology, therefore, has facilitated the practice of network governance in real time.

Multinational organizations have developed different models of network governance depending on whether their dispersed sites are equal partners, dependent satellites, tentacles of the headquarters, or the result of mergers. In this context, multinationals have established different models of corporate network governance that best suit their operation in a competitive, international, free-market environment.[9]

In a sense, one could argue that the corporate network governance model has been borrowed from the practice of the global administration of the Catholic Church. Long before the advent of multinational corporations, the Catholic Church has been involved in corporate network governance, with the pope at the helm in collaboration with Church leaders in other parts of the world.[10] The structure of this global network form of governance has evolved over time with periods of expansion. Since Vatican II, there has been a period of stability characterized by the fine-tuning of the relationship between the dioceses and the bishop of Rome.

What the network concept adds to this equation is the transnational/global link, the mutual dependence of the nodes, the shift from the governance of a node to that of a larger international circuit, and the legitimizing of the center by the periphery and vice versa.

In corporate network governance, networking is done through a corporate structure and the actors are located in more than one single place. Although the corporate structure of each node may be organized hierarchically, the network itself is not necessarily organized in a hierarchical manner. Each node is seen as an equal partner and has no restrictions in comparison with others in terms of what it can contribute to the network. All the nodes contribute according to their own capacity to the governance of the network. The leader of any node may be elected to serve as the head of the global network of Catholic churches. In the Catholic Church, thus, network and hierarchy complement each other in the performance of the global governance of the ensemble.

THE HOLY SEE IN THE VATICAN AS A CENTRAL NODE

The Vatican, as the central headquarters of the Catholic Church, is also a node in the global governance of the Church, but because of its primacy inherited from and attached to Saint Peter and Saint Peter's tomb, it has precedence over other nodes (Canon Law article 330). The Vatican governs as a central node of a global network, and as such, the pope and the Curia are responsible for the everyday management of the current affairs of the universal church. Each unit in the network is responsible for the resolution of internal problems, following the canonical principle of subsidiarity.[11] According to this principle, some decisions are made locally, whereas others are made by the Vatican on behalf of the network. Intranode problems are settled locally, but internode problems call for higher authorities to intervene for their resolution.

The principle of subsidiarity implies that the headquarters may delegate its decision-making power to another node in the network to resolve a problem in that node or a conflict that may affect the rest of the network.[12] Subsidiarity implies a division of labor, which recognizes the role of nodes in the governance of the network. In such a system of network governance, a central node makes certain decisions, and subaltern units may be called upon to make others. This balancing act is a way to facilitate and entertain the collaboration of nodes in the governance of the network and to ensure that their input plays a role in the resolution of an issue. For example, if the Latin Patriarch of Jerusalem were left to make all the decisions concerning the holy lands, it is likely that these decisions would tilt more in favor of local interests than toward the interests of the global network. Similarly, if the Vatican were asked to make all decisions concerning the holy lands without consulting the Patriarch, policies would tilt more toward the interests of the global network than those of the local Christian community. To maintain a healthy balance so that the interests of both parties can be represented, these two governing bodies engage in dialogue with each other. Even with the corporate leadership provided by the Vatican, network governance is still the backbone of its mode of operation.

INFRASTRUCTURE OF NETWORK GOVERNANCE

The deployment of the globalization process implies that an appropriate transnational infrastructure exists to sustain ongoing cross-border activities and traffic flow. Infrastructure is understood here in terms of physical structure, social institutions, territorial units, and transnational associations. Infrastructure provides a basis for the production and absorption of global flows of information, people, images, and goods. In this sense, the infrastructure is constitutive of the globalization process itself to the extent that it can shape its content, influence its direction, and perturb its rhythm and cadence.

The Catholic Church has developed infrastructural signposts throughout the globe as a result of the missionary system that fuels its geographical expansion. Local churches, which are units of the global network, anchor global Catholicism in specific national territories.[13] These are not exclusively national churches, but units that use their global linkages to define their modes of national belonging. Each is a global church that choreographs its practices in a national context. In this context, the global identity constrains and shapes the performance of the local identity.

The universities and research centers that the Catholic Church has established throughout the globe have also contributed to sustaining globalization. They do so by facilitating the foreign exchange of students and faculty, by encouraging or sponsoring participation in international scholarly conferences, and by producing scholarship of discovery for the consumption of both Catholic and non-Catholic readers. For example, Catholic interpretations developed in one country may be consumed and absorbed by Catholic academics in other countries.

Conferences of the superiors of major religious orders, or provincials of the same religious order, as well as transnational lay Catholic associations which focus on religious life, social justice, and any other public-policy issues, are organizations that occasionally meet in global forums to discuss matters that concern their international participants. These organizations have existed for years and provide their members with a platform for discussion, global solidarity, and a deepened understanding about the requirements of their faith.

Catholic Christianity thus is doing more than simply making an effort to be global. It is doing more than just adjusting itself to the globalization process. It is doing more than merely attempting to deterritorialize itself to avoid state constraints. Instead, Catholic Christianity has effectively established a global infrastructure in order to enhance the performance of its global agenda.

Furthermore, the development of the global infrastructure is not exclusively the work of the Holy See. This infrastructure is also developed by other units of the global network. In this chapter, we will see that various units, with the collaboration of the network, entertain transnationalization. Each type of social infrastructure of globalization becomes totally integrated because of its incubating role, its ability to feed the needs of a cross-section of the network, and its power to enhance the viability of the network.

The institutional bureaucracy that governs global Catholicism relies heavily on transnational agencies to carry out its everyday work. These agencies are located not only in the Vatican and throughout Rome, but also around the globe at diocesan chancelleries, national and regional bishops' conferences, the diplomatic posts of apostolic nuncios, and religious orders' national headquarters. These agencies either interact continuously with overseas associates by providing guidance, developing policy, solving problems, or assessing progress reports submitted from abroad to sustain their collaboration with the rest of the global network.

In these agencies that constitute the administrative backbone of global Catholicism, one finds a number of different kinds of actors: *global bureaucrats*, that is, individuals whose foremost task is to provide leadership and, when necessary, to resolve network problems generated by local nodes; *transnational bureaucrats*, who travel to places to fix problems, serve as intermediaries to the Vatican concerning the worries of a node or vice versa, or whose role is to interface the local with the rest of the global network; and *local bureaucrats*, individuals whose task is to resolve local problems in light of norms established by the global network through the Holy See.

In the same local site, one may encounter plural niches with specialized agents that engender distinct global flows. These agencies specialize in solution transfers, and because they deal with different sites, a solution found in one may be used to solve a similar issue elsewhere. The feedback they receive nationally or from overseas allows them to adjust their mode of operation and shows how nodes interact and influence each other on a global scale.

There are different transnational platforms that feed and sustain the performances of global Catholicism. For example, one must distinguish transnational religious orders from transnational homeland congregations and diasporic congregations from sister-church congregations.[14]

Most Catholic religious orders, such as the Jesuits, Augustinians, and Carmelites, were established with a global vocation and outlook. Not only are they open to citizens of any country, but they also have mission churches outside their national birthplace. They evolve as global bodies with a headquarters to oversee the activities of all the branches and to ensure that the network sustains itself. The branches maintain relative autonomy, despite their dependence on the order's international headquarters for final approval of any decision or proposed activity. For example, the international headquarters must approve the election or selection of a regional provincial.

The transnational homeland congregation is a religious community with a core base of residents and a floating base of diasporic members who, from time to time, return from abroad and actively participate in the affairs of the congregation.[15] These diasporans use the congregation for basic rites of passage, such as baptism, marriage, and funerals, and may also provide financial assistance to their homeland parish.

The diasporic congregation is a host-land-based group or a diasporic church. The members live in the diaspora, but continue to help the homeland, which includes their respective homeland congregations. They might raise funds for the homeland, invite their former priests and bishops for a visit, have pastoral letters from the homeland read to them during their

Sunday services, or even request to borrow a priest from the homeland to minister to them.

In regards to the transnational sister-church congregations, these are often churches in the Global North that link up with churches in the Global South for the purpose of sharing financial and material resources and engaging in common projects with them. For this to materialize, they exchange visitors, discuss what needs to be done, raise funds in the parishes of the Global North on behalf of sister parishes in the Global South, recruit volunteers, and collaborate with each other. This is done under the radar of the Catholic Church, with the full knowledge and consent of the local bishops and the implicit approval of the Holy See through the nuncio. A local church in the Global South may serve as a sister-church to more than one parish in the Global North and vice versa.

SPATIAL MATERIALITY OF CORPORATE NETWORK GOVERNANCE

The spatial materiality of network governance can be unveiled by identifying and analyzing the territory-based and non-territory-based circumscriptions of global Catholicism. Because Catholicism develops its own spatial units—not all of which coincide with the territorial divisions of the nation-state—it is important to identify them in order to see how they fit in the network governance scheme. One may a priori presume that the map of the globe is not identical to the map of global Catholicism. In fact, while the globe is dissected into nation-states governed by different leaders with no state in charge of the total landscape, Catholicism, in contrast, has a global spiritual leader responsible for the good governance of the entire network. Its territorial and administrative units form one single, but differentiated, global religious landscape.

The governance structure of global Catholicism, as spelled out in various articles of the canon law, proceeds from territorial and non-territorial circumscriptions as its base of operation.[16] There are several units in this structure: for example, the diocese is headed by a bishop appointed by the Roman pontiff and the bishop, in turn, belongs to the college of bishops responsible for the corporate governance of the Catholic Church. The territorial prelature, for its part, is a unit headed by a lower-rank clergy or a "bishop prelate" (the Prelature of Montigny in France exemplifies well this spatial category). The territorial abbey is headed by an abbot of a nearby monastery, and both the secular and the regular clergy that minister to the brethren fall under the aegis of this abbot. The mission diocese is an established circumscription whose leadership is entrusted to a religious order, such as the Jesuits, Dominicans, or Franciscans. The apostolic vicariate is also a mission area headed by a vicar apostolic, usually a bishop appointed for this special post by the pope. The apostolic prefecture is a territorial unit headed by a prefect apostolic (not a bishop, but a member of a religious order established in the area).[17] Although this list does not pretend to be exhaustive, we mention these territorial circumscriptions not only to indicate their relations to the Holy See and the rest of the network, but also because they are the most common administrative units in the Catholic Church. They depend on different congregations of the Curia in Vatican City for guidance, financial assistance, and appointment. For example, some territorial abbeys fall under the jurisdiction of the Congregation for the Bishops, whereas others fall under the jurisdiction of the Congregation for the Evangelization of Peoples, and still others are supervised by the Congregation for the Oriental Churches.[18]

Not all ecclesiastical circumscriptions are based on territorial divisions. Some evolved on the basis of other criteria. In this vein, one can think of the Military Ordinariate, which cares for the members of the armed forces of a particular country, or the Latin Ordinariate for the Faithful of the Eastern Rite, which ministers to the needs of Catholics of the Eastern tradition who live in Western Europe or North America.[19] This multiplicity of spatial configurations—whether in the form of missionary churches, incomplete institutions that have yet to attain maturity, or based on territory or types of apostolate, different rites, or policies of the congregations of the Roman Curia—has come about by reason of history and tradition. The hierarchical status of these circumscriptions in the pecking order set forth by canon law indicates the volume and direction of communicational traffic flows, their ability to influence the network, and the intensity with which the network is able to influence them.

The boundaries of ecclesiastical circumscriptions evolve over time to reflect demographic change and the needs of the population. Redistricting may occur, in which case a diocese splits into two and establishes an additional, contiguous diocese. Similarly, the boundaries of a diocese may expand to include an additional area. There may also occur an evolution in the life cycle of an ecclesiastical circumscription: it may start as an apostolic prefecture and then transform into a diocese. A diocese splits in two often because its population explodes. In contrast, two dioceses may combine because of dwindling membership, lack of resources, unavailability of priests to staff parishes, or for other reasons.

This interpretation supports the observation that units are not equal in status and that networks are composed of nodes of unequal status. Corporate network governance is a modulated form of governance that upholds these differences between nodes, facilitates the transition of some to a higher status, and attends to their different needs. So the hierarchical system affects the network system. In this instance, the network sustains and legitimates the hierarchy, and the hierarchy shapes the transnational order of the network. The hierarchy influences the network through the implementation of the canon law, which legislates forms of connections to the Holy See and validates the activities of every unit.

The differences between units can also be measured by territorial size. For example, all dioceses, with the exception of two types, exist alongside other dioceses in the same nation-state. They exist nationally as a network of dioceses. However, there are two cases that do not fit this model. One can think of a diocese that covers the territory of an entire nation, such as Monaco, or that of an entire island, as in the case of Gibraltar. Here, again, these nodes participate in the global circuit with their own peculiarities and local issues.

THE GLOBAL MOBILITY OF CLERGY

Given that global movement may involve the withdrawal of the formal status of a member of the clergy in one diocese (disincardination) and reincardination in another, incur expenses to be paid by a local church, and affect the availability of sacramental services to a parish or the ability of a clergy member to continue living within the framework of canonical rules, the Holy See understandably monitors the mobility of clergy from one country to another (Canon Law articles 265–267).[20] Catholic clergy travel internationally for diverse reasons. Some travel to overseas appointments, as in the case of a bishop named to head a foreign diocese or to occupy a post in the Curia. Some travel for postgraduate theological training, are assigned to care for migrants from their home country who now live abroad, or become Vatican diplomats or delegates. Still others travel to attend regional meetings or to make the canonical visit to Rome every five years, a requirement bishops must fulfill. Because such mobility intrinsically involves using Church money, accountability to higher authorities is deemed necessary to prevent abuses and mismanagement of funds.

This mobility is necessary to help communities in need, wherever they are located, as well as to pursue the goal of the Church to preach the gospel at home and abroad.[21]Because the mission of the Church is to be present everywhere, it assigns missionaries to serve the communities where it has established congregations of practitioners. Mobility is also used to resolve problems that cannot be solved locally, to discuss issues of concern to more than one node, and to provide global governance to the flock.

More recently, the Holy See has developed policies and procedures concerning the migration of priests from the Global South to the European Union and North America. These are priests who migrate for further training and decide to stay abroad, or who ask to be incardinated in a foreign diocese. The overseas migration of priests is not a recent phenomenon. Earlier, they migrated from North to South and from Europe to the rest of the world, including the United States, for missionary purposes. The migration of Latin American, Caribbean, Asian, and African priests to North America and Europe, often to care for diasporic communities from their homelands or to relieve a shortage of priests in these industrial countries, is a more recent occurrence. In the United States, this is made possible by the shrinking number of home-grown priests. These foreign clergy often are hired to take care of soon-to-be-abandoned parishes. Some do not return to their countries due to the economic incentives (which act as magnets) in developed countries. The Holy See has published new norms to govern this nascent form of clerical mobility.

In the nineteenth century, priests were recruited in Italy, Poland, Ireland, and France to care for diasporic communities in the United States.[22] One can think of Polish priests in Chicago, Irish priests in New York, French priests in Baltimore, and Italian priests in Boston who provided services to enclaves where compatriots congregated. Now that Catholics migrate en masse from Latin America to the United States, the Church in the latter region recruits priests from the former to serve these immigrant communities.[23] To care for these dispersed communities, some bishops have invited foreign priests from Latin America and the Caribbean to join their diocese. This is often needed because of the scarcity of priests who are familiar with more than one language. Appointing priests from the homeland to provide this service may in fact serve as the best solution for closing the gap. Sometimes these priests are on loan for a number of years or serve on a contractual engagement, other times they are permanently transferred to the overseas diocese if they have asked for incardination.[24]

For the sending diocese, this diaspora-driven mobility has both disadvantages and advantages. The disadvantage is that a poor diocese, which paid for its priests' training and expected to assign them to local parishes, can no longer rely on their services, which are now being provided to another country. The advantage is that the priests may serve as a vehicle to channel special diasporic collections to the homeland. The consequence of this cross-border phenomenon is that some homeland bishops consider the diaspora to be an extension, or rather overseas members of their dioceses, and in some cases, even send pastoral letters to be read to these diasporic congregations or visit them for fundraising purposes.

Missionary work, which entails mobility from one part of the world to another, remains a vital component of Church activities. Religious orders are always engaged in some form of mobility to staff their own missions. By and large, canon law, the Curia's policies, and general monitoring by the Holy See calibrate the governance of the clergy's mobility. The Church moves clergy around to meet its organizational needs and to serve the global Catholic community in the most efficient way possible.

The mobility of clergy has led to the formation of what might be called a *clergy diaspora* in some countries. Such a diaspora has its own characteristics in terms of its relations with the homeland and the hostland. It is an elite diaspora because it occupies leadership positions in the hostland; engages in transnational communications with the homeland socially and institutionally for canonical reasons, also for the purpose of fundraising and remaining in contact with family members and friends; maintains sustained links with compatriot-owned businesses in the hostland as well as the homeland embassy as potential sources of funding for local church initiatives; and, in some cases, forms its own associations—such as the Fraternité Saint Jacques—to which the members of the French missionary clergy in Haiti belong. The relations of such associations with the local bishops add one more feature to the network governance of local nodes.

The issue of priestly migration or clerical diaspora is a network problem—not simply the problem of a single node—and therefore can only be resolved through network governance. It concerns the well-being of the priest, the homeland congregation left behind, the bishop of his diocese, the new congregation that he is serving abroad, the bishop in the receiving country, and the Vatican bureau charged with overseeing such mobility. Questions that arise include: Does the priest play a missionary role? Is this a network response to the shortage of priests in the United States and the European Union? Or is he there only to care for his compatriots? Other factors to take into account are that he could be assigned to a foreign post by his bishop, recruited by the bishop of the receiving country, or appointed to a position abroad by the Vatican. Each of these aspects concerns the well-being of the global church despite the local colorations involved.

TRAINING FOR GLOBAL LEADERSHIP

The Catholic Church is involved in bringing seminarians to Rome for basic indoctrination, which is the orthodox training in Church doctrine and canon law. These seminarians constitute three groups: college students, newly ordained priests, and members of religious orders. College students are selected to study for the priesthood in one of the Roman colleges (the North American College, the French Seminary in Rome, etc.) or universities (Gregorian, Angelicum, etc.). They attend one of the Roman colleges sponsored by their country's bishop, as in the case of the French Seminary in Rome or the North American College. Newly ordained priests may also be selected for postgraduate training, often in canon law. And of course, religious orders such as the Jesuits and Dominicans routinely send students to the Pontifical Gregorian University (for the Jesuits) or the Angelicum (the Pontifical University of Saint Thomas Aquinas, for the Dominicans), both in Rome, which provide a pool from which professors for the universities they sponsor throughout the globe will be selected.

Students' training is funded by the diocese, a scholarship from the university, or, in some cases, by a humanitarian foundation. Seminarians cannot gain admission to a Roman college by themselves, but must be selected or approved by their local bishop after they have proven themselves based on their academic scores and attachment to their vocation.

The foremost goal of this foreign training is to prepare the local cadre, those who most likely will serve as bishops. It is an investment by the local church for future service at a higher level. The fundamental goal is for the seminarians, through their exposure to the Church bureaucracy, to acquire a deeper knowledge of Church beliefs and laws and some Italian to facilitate eventual communication with the Holy See. The secondary goal is to ensure that congregations outside of Rome subscribe to the doctrine propagated by the Holy See in terms of dogmatic theology. This practice is undertaken to reduce the possibility that regional traditions out of sync with Roman Christianity will develop. Furthermore, it prevents postgraduate clergy from reinventing the wheel and from developing their own school of thought at variance with the Holy See, but it also trains clergy to teach the orthodox path of Roman Catholicism. Because the Vatican wishes to emphasize its multicultural makeup, it prepares clergy who are selected to study in Rome to lead their national churches at the episcopal level while also indirectly grooming them for potential special assignments in the Vatican's institutions.[25] More recently, a few seminarians—mostly from Africa—have been selected to study in Rome in order to join the Vatican diplomatic corps, a direct and efficient route to join the Church hierarchy and acquire the title of monsignor, bishop, or even archbishop.

LOCAL COLLECTIONS FOR GLOBAL REDISTRIBUTION

The Catholic Church is the only religion that globalizes and centralizes the mechanism by which it collects and distributes money. With regard to how economic policies are formulated and administered in the corporate network governance of the Catholic Church, the focus here will not be on how global Catholicism generates its income, but rather on how it shares resources among its various constituencies. The mechanism that triggers the movement of funds from one Catholic unit to another and the governance oversight involved are the objects of our analysis.

That said, we will briefly note that with regard to its economic policies, the Catholic Church's financial situation is complex because the institution also runs an independent city-state, which has obligations pertaining to its civil status.[26] As a result, the Vatican has its own financial issues running the affairs of the state, in terms of revenues for services rendered (entrance fees paid to visit the Sistine Chapel and the museums) and expenses to cover employee salaries, finance diplomatic missions, or defray the cost of the pope's travel abroad. Other sources of state revenue include the Vatican Bank—that is, the Istituto per le Opere di Religione, created by Pope Pius XII in 1942—and money invested in stocks. Parishes, religious orders, and some lay Catholics do invest in the Vatican Bank, and any dividends it makes, whether from low-interest loans to Catholic institutions or from interest earned on savings, are mostly used to finance missionary activities and to help dioceses in financial distress. It is a financial institution with exclusively humanitarian and religious purposes.

Viewed in terms of the network arrangement that allows the global collection and redistribution of funds, it is important to note that, however dispersed local congregations are, the Catholic Church constitutes one body, and its resources belong to all the nodes. In other words, the Holy See coordinates its global collection and redistribution of money to be shared for the good of all the brethren. Viewed in those terms, the papacy functions as a mechanism of redistribution among Catholics.

When it is claimed that the papacy functions as a mechanism of redistribution among Catholics, it means that it takes the money from the rich and gives it to the poor. Peter's Pence, a collection taken in every parish once a year, and any special appeals by the pope, for example, provide the Holy See with funds for global charitable distribution. A portion of this money is given to communities experiencing acute need, and some of it is invested in order to respond to future natural disasters, support missionary work, or help alleviate poverty in general. The size of the collection varies from country to country and from parish to parish. Christian communities in the United States, Canada, and Germany lead the pack in terms of the amount of money they annually contribute to these funds. This mechanism has the same rationale as the welfare system run by nation-states. In the state model, the money is collected through taxes, and while in the Church, it is a voluntary contribution, but the goal—to help communities experiencing some form of distress—is the same.

The transfer of funds across global networks is done to alleviate the burdens that the headquarters places on the nodes, to help specific units in the network, to respond to a request from a node or from the Vatican on behalf of a node, to undertake the missionary projects of the religion, and to keep up the physical and social infrastructure of global religion.

The governance of the Church budget is a collective and global enterprise. This is seen in the appointment of a CEO of the Vatican Bank, who, during the papacy of John Paul II, was an American archbishop. In 1979, the cardinals convened to discuss Church finances and to provide guidance for reducing the deficit and proposing new guidelines for the management of Vatican funds. From time to time, local churches are called upon to help the Holy See solve deficient budgets at the Vatican.

Global participation and interaction of units in the network is also seen in the acquisition and sale of any Church property that is worth more than one million dollars. The issue must be presented to the Holy See for approval. This indicates that Church property belongs to the universal church, and not to the local churches (Canon Law, article 1273). The local church is the custodian, not the owner of the property. To protect Church assets in case of insolvency, any local church that received a substantive loan from a bank (relative to the country, of course) must get approval from the Vatican before closing the deal.

The redistribution mechanism follows formal established routes, some through the corresponding office in the Curia, such as the Congregation for Seminaries, which may sponsor the study of a seminarian in Rome, or Cor Unum, whose mission is to help relieve poverty in poor Catholic communities. Sometimes the Holy See is not directly involved, but rather sponsors or encourages redistribution as in the case of the Middle Eastern Foundation in New York or Caritas in Germany. But more commonly, the pope distributes monetary gifts to clergy visiting from poor countries for ongoing projects in their dioceses.

By focusing on how nodes in the network generate income to defray their expenses, one can see the network's financial situation. Administratively, dioceses are autonomous, keeping large portions of the money they collect from parishioners as well as the returns on their investments. The sum of money that dioceses distribute is negotiable, depending on the amount they collect, state laws, and the expenses they have to cover. The Vatican has the same level of autonomy and rescues other units depending on what it can afford. The rule of thumb here is that each unit is called upon to finance itself. However, if it is not able to do so because of poverty, natural disaster, or war, other units are expected to help to maintain the health and stability of the node and hence of the network.

Within the global network, funds are moved in a continual and routine manner as a result of institutional or canonical obligations. They are dispersed from one country to another. For example, one can look to the Congregation for Seminaries and Educational Institutions (Curia), which finances some missionary activities, and l'Oeuvre de St. Pierre L'Apôtre, which subsidizes some seminaries to train future priests. Church money moves transnationally in three directions: from the Vatican to struggling local dioceses, and vice versa, and from dioceses in the Global North to dioceses in the Global South.

GLOBAL SECURITY FOR A GLOBAL CHURCH

Catholicism, Judaism, and Islam each have unique ways of ensuring their security. The practitioners of Judaism have the most complete and developed security system to protect their synagogues and worshippers. The everyday security operations of Catholicism have a more local, rather than global, focus.

Established in 1506, the Swiss Guard protects the pope and assures security inside the Vatican city-state. The role of the guards is not very different from that of any police force. In that capacity, for example, they check people who come through the city gates, and they have their own intelligence operations to find out what is occurring on the Vatican grounds. The Swiss Guards were originally created as a purely local institution, and have remained so. In 1970, an additional police force was established called the Corpo della Gendarmeria dello Stato della Citta del Vaticano (Gendarmerie of the Vatican City State). This new labor division assigns the protection of the pope to the Swiss Guard and the policing of the Vatican to the Gendarmerie.[27] Both security forces coordinate their activities with the Roman police, the State Police, and the Italian National Police, not only to maintain order during events that attract large groups of pilgrims, but also to assure the pope's safety when he travels outside Vatican City walls.

The different levels of threat experienced by each religion globally affect the ways in which the three faiths manage matters of security. By far, Judaism is the most strained because of the long history of persecution against religious Jews. Catholicism has not had to deal with this issue as an ongoing problem and concern. Persecution, when it happens to Catholics, tends to occur in a specific country and may last for a short period of time. By and large, tolerance seems to be the general attitude of society toward this religion.

THE HOLY SEE AND THE HOLY SITES

The governance of holy sites is one aspect of the global leadership provided by the network of Catholic practitioners which encompasses both the laity and the clergy. Each holy site constitutes a global node because of the role of the network in its administration, the pilgrims it attracts from around the world, the money it collects that must be shared with the Holy See as well as the money the Holy See may use for its maintenance, and the diplomatic relations that nuncios may undertake for the protection of such sites. In this context, corporate network governance entails the governance of the holy sites that are under the direct control of the Holy See and negotiations with foreign governments for the protection and access of other sites. Corporate network governance concerns three types of sites: those that are located in Rome for which the Holy See has direct custodial responsibility, those that are under the supervision of a national church or religious order, and finally, those that are in the holy lands of Israel and Palestine.

The Roman sites include both the tombs of the early apostles, such as Peter and Paul, and those of prominent saints, such as Francis of Assisi. The Holy See is responsible for the maintenance and protection of sites located inside the Vatican, for access to them, and for the collection of entrance fees to pay for infrastructure, personnel, and security expenses. These sites are often shrines visited by pilgrims who are provided with literature to enhance their understanding of the lives of saints. The shrines are used for personal piety, but also as didactic tools to teach about the life of the Church. They are also the objects of canonical rules, whereby regulations are set to ensure that overseas bishops, during their *ad limina* journey to Rome every five years, visit the tomb of Peter to strengthen their ties to the Holy See.

The governance of holy sites under the control of a subsidiary church (Nuestra Senora de Guadalupe in Mexico, for example) or a religious order (such as Notre Dame du Cap de la Madeleine in French Canada) is left to the charge of this subsidiary entity, but with rules and regulations set by the Holy See. Canon law and special policies formulated by the Curia govern such sites.

The governance of the holy sites in the holy lands has varied over time, depending on whether the Catholic Church (during the period of the Crusades), the Muslim authorities (between the mid-1300s and mid-1900s), or the state of Israel (post-1967) directly supervised the security arrangement to protect them. Despite the fact that international law has not yet settled the status of Jerusalem, because the holy sites are located in Israel and the Occupied Territories of Palestine, the Holy See has insisted on the right to protect its patrimony, maintain easy access to the sites by Christian pilgrims, and pursue religious activities as it sees fit.

In practical terms, the Holy See establishes requirements for those who are employed to work at these sites. These regulations range from dress code to marital life. For example, a divorcee is not allowed to be a Vatican City employee. Broad guidelines are provided to local churches to be implemented at their discretion. The Holy See also encourages the faithful to come to Rome as pilgrims, which they may do in the company of their local priests, who serve as their guides.

The protection of the patrimony is a central aspect of network governance for Catholic Christianity. A special office of the Curia is in charge of administrating these sites. While the Holy See coordinates the global governance of these sites, the entire Church engages in financing their maintenance, which is required to keep them accessible. For this purpose, every year, a collection is taken on Good Friday in every local parish for the maintenance of the holy sites in Israel and the Occupied Territories.

Corporate network governance in the Catholic Church thus has an administrative aspect, which involves both the Curia and the local church in the management of resources; a rule-making aspect, in which rules are established by the Curia, but affect pilgrims and local administrators; a financial aspect, with everyone called upon once a year to donate money for the upkeep of the holy sites; a patrimonial aspect, apparent because the property belongs to the universal church—not the local church—and the sale of land or buildings requires approval by the Holy See; and finally, a doctrinal/ritual aspect, because any teaching or liturgy undertaken must be in line with the official teaching of the universal church.

OUTSOURCING A GOVERNANCE FUNCTION

The management of a corporate headquarters that provides leadership to a global network entails decisions concerning decentralization, takeovers, mergers, and outsourcing. In this light, the administration of the Catholic Church confronts issues similar to the running of any major firm or multinational organization. Although maintaining this governance is usually the prerogative of the headquarters, sometimes the function is outsourced to another node. This way the global network can be seen as multi-headed and some functions can be delegated to other nodes.

Administrative decentralization is effected to allow the headquarters to concentrate its efforts on other matters and may empower a unit that is better equipped to run the agency. In the functioning of corporate network governance, an administrative function can be outsourced to a node whereas the power to make final decisions remains with the headquarters. It is segmented decentralization, because only the administrative aspect is affected.

The Catholic Near East Welfare Association (CNEWA) is a good example of outsourcing an administrative governance function (www.cnewa.org/default.aspx). Created in 1926 by Pius XI, it was outsourced to the cardinal archbishop of New York. Although its central office is located in Vatican City, the headquarters and president of the association are both in New York. The Association maintains regional field offices in Jerusalem, Amman, Beirut, Addis Ababa, Asmara, and Ernakulam. The beneficiaries of its services and support are the peoples of Israel and the Occupied Territories of Palestine, Egypt, Syria, Lebanon, Jordan, Iraq, Ethiopia, Eritrea, and India. More recently, it has expanded its field operations to include the Orthodox churches of Eastern Europe, including their diaspora churches in the Americas.[28] The heavy-duty financial and administrative work is done by the headquarters in New York, with the Holy See providing oversight.

The work of the Catholic Near East Welfare Association is not confined to the Latin Church or to Christians.[29] Aid is provided where it is most needed. In 2003, it helped financially the renovation of parish churches in the Maronite Eparchy of Saida (Lebanon), contributed to the completion of an outreach clinic for the Caritas Baby Hospital in Nahaleen (Palestine), supported the maternity Hospital Al Hayat in Bagdad (Iraq),and paid for the repair and renovation of the Melkite Greek Catholic Church in Kerak (Jordan). It contributed to the expansion of St. Nektarios Greek Orthodox Church's youth activities program in Amman (Jordan), to the renovation of the Armenian Catholic church in Damascus (Syria), and to the administration of a support services center for homeless mothers and their children in Addis Ababa (Ethiopia).[30] It also provided emergency assistance to Sudanese refugees in Cairo, as well as scholarships to seminarians at the Coptic Catholic Patriarchal Seminary of St. Leo the Great in Egypt.

The CNEWA is the site where three networks—service, financial, and leadership—meet, crisscross, and link up with the New York headquarters, which in turn connects their nodes in order to be able to rescue specific units when necessary. Each network contributes to the success of the operation.

The service network is its principal raison d'être. It primarily confines its services, as the name suggests, to the people and churches in the Middle

East and those of the Eastern Rite, located outside the region in South Asia (India) and Eastern Europe. This network links donors and administrators to the recipients of these services.

The leadership network sheds light on a different geography of operation of the association. Here, the linkages are between the Vatican and the trustees of the Association—the archbishops of New York, Quebec (Canada), Baltimore, Ottawa (Canada), and Seattle. These trustees are seconded by other Canadian and American bishops who serve either as officers or directors in the New York headquarters and the Ottawa branch office.

The financial network brings to the table still another set of players. They are the major donors: Catholic Relief Services (United States), USAID (United States), the Bischofliches Hilfswerk Misereor (Germany), the Internationales Katholisches Missionswerk (Netherlands), the Archdiocese of Cologne (Germany), Caritas Switzerland (Switzerland), and many others.

The major lesson learned from focusing on the operation of a node such as this is that network governance is made up of more than one network, and one must identify its crisscrossing pattern to understand the relationships between networks, such as how a network may contribute to the existence of another and how a node may be able to help another. Whenever a node is called upon to play a leadership role vis-à-vis the circuit or another node, it is likely that there will be a differentiation of networks based on the services rendered, the collection of funds, and the leadership environment. The success of the operation depends on the harmonization of these three sites of network crossing: the collaboration among leaders to provide guidance, the generosity of donors to sustain the effort, and the existence of a need to overcome any problems and harmonize the life of a node with the rest of the network.

IT AND NETWORK GOVERNANCE

The use of IT by the Church leadership at the diocesan level and by nuncios and delegates on missions abroad has become a central component of the corporate network governance of Catholic Christianity. The Vatican sees IT not only as a means for efficient communication to run a worldwide bureaucracy, but also as a tool that can be exploited to propagate the Christian message and to modernize the modes of the Church's operation. Corporate network governance, in this context, requires the ability to develop a policy framework discerning the morally acceptable or unacceptable use of IT in human affairs. The Church appropriates and exerts the right to guide the moral behaviors of the brethren and has done so through official public statements from the pope and from the office in charge of telecommunications in Vatican City.

Since the Gutenberg revolution, the Holy See has used mass media to advance the cause of the Church.[31] Already in 1587, Pope Sixtus V established LibreriaEditriceVaticana, the official printing press that is still used to publish religious literature and papal documents.[32] The Libreria also prints papal encyclicals, Curia documents, other Catholic literature and the *Osservatore Romano*, the official newspaper that disseminates news about the Catholic community and papal initiatives. Through this Vatican-controlled newspaper, the clergy remains abreast of new developments in Rome and throughout the globe.

The Vatican has been using the telephone for as long as the rest of civil society. At the inaugural ceremony marking the transition from the old manual switchboard to an automatic dial tone in 1930, Pope Pius XI said, "It gives us great pleasure to inaugurate in the Vatican State this new automatic system which represents the height of perfection in telephony, and which will offer this Holy City a service worthy of the See of the Vicar of Christ."[33] This is the telephone service used by the pope and his entourage to communicate with the Catholic diplomatic missions abroad. The cell phone has more recently been introduced into the Vatican to complement the land-based telephone system.

Radio Vaticana, staffed and administered by Jesuits, has been an important medium through which the Vatican has ministered to the global Catholic community in general and those who live in Central and East Europe, in particular.[34] With its shortwave and medium wave broadcasts in more than thirty languages, it reaches listeners in more than one hundred countries and in such faraway places as Haiti, Brazil, and Japan.[35]

Since 1997, the Holy See has had its own Internet Office that provides information through its website (www.vatican.va) to the Catholic communities in Italy and abroad.[36] Here one may learn about the history of the papacy, find links to any of the offices of Vatican City, search for the pope's latest public pronouncements, access the contents of the most recent encyclicals, or learn about the visitors or pilgrims received by the pope in public audience. The projected global readership of the website is manifest in the different languages (French, Spanish, English, German, etc.) used to make information available.

Pope John Paul II identified a number of ways in which the Internet could be of use to the Church. He saw that the primary benefit was "to proclaim the Gospel message" and identified a new field of apostolate, that of "helping those who first make contact through the Internet to move from the virtual world of cyberspace to the real world of Christian community."[37] Recognizing that the Internet makes useful information on the Catholic Church available to believers and non-believers alike, the Pope noted that, "while the Internet can never replace that profound experience of God which only the living, liturgical and sacramental life of the Church can offer, it can certainly provide a unique supplement and support in both preparing for the encounter with Christ in community, and sustaining the new believer in the journey of faith which then begins."[38]

IT has been used in very creative ways by the Church. Nicola Clark informs us that "Wi-Fi laptops and personal digital assistants" from

Motorola are used by the Swiss Guard to protect the pope and provide security inside the Vatican and that "the transmission of the pontiff's prayers and teachings as text messages to mobile phones . . . allows subscribers to receive Pope John Paul II's Thought of the Day as an SMS, or short message service" in Italy, Ireland, Britain, Malta and the United States.[39]

The effective use of the television and the Internet by John Paul II, who established the Centro Televisivo Vaticano (Vatican Television Center)[40] in 1983, set the pace for his successor Benedict XVI. This Centro produces films about public activities at the Vatican, such as the Pope's General Audience with pilgrims, politicians, or members of civic organizations; his trips abroad; and public religious ceremonies, such as the canonization of a saint. It broadcasts these events directly from the Vatican and makes the films available to Catholic and non-Catholic TV stations around the world, as well as archives them at the Vatican for future reference (http://www.vatican.va/news_services/television/index_it.htm). CTV showcases events daily through its video newscasts and weekly through the production of its popular program "Octava Dies."

The Vatican sees the public that it wants to reach through the Internet as divided into two groups: the *digital generation*, those who have grown up with the Internet and have begun to use it at an early age; and the *digital immigrants*, those who were familiar with older modes of communication, but have developed a taste for the Internet and begun to use it in their midlife or later. In the view of the Vatican, these groups have different needs and therefore require a different strategy of action to reach them.[41]

In a continuing effort to inform the Catholic community around the globe in general and the digital generation in particular, in 2009 the Vatican established the Vatican Channel on YouTube (www.youtube.com/user/vatican). Through this video-sharing website, the Vatican hopes to reach a larger public beyond its membership and to have some control over the way it presents the activities of the Holy See to the Internet audience. Also in 2009, the Vatican established another presence in cyberspace through its Facebook page. With an iPhone, iPod, or any smart phone, or any kind of Internet Connection anyone around the world can now access directly from the Vatican information about the daily activities of the Holy See (available in Italian, English, French, German, Spanish, Portuguese, Hungarian and Chinese), or download the book of daily prayers for priests with the iBreviary application. The portal Pope2You gives access to the Pope's speeches and his travels.

POLICING THE BORDERS OF CATHOLIC CHRISTIANITY

The Holy See has developed an elaborate structure to police the doctrinal, ritual, and sacramental borders of the Church. The Congregation for the Doctrine of the Faith, the Congregation for the Oriental Churches, the Pontifical Council for Promoting Christian Unity, the Pontifical Council for

Inter-Religious Dialogue, the Commission for Religious Relations with the Jews, the Secretariat for Non-Christians, the Secretariat for Non-Believers, and, more recently, the Interdicasterial Commission for the Church in Eastern Europe (established by John Paul II on January 15, 1993) are the official organs of the Vatican that are responsible for providing guidance and policing the formal interface of the Catholic Church with other faiths.

In its relations with other religions, the posture of the Catholic Church has evolved over time from that of evangelization and conversion of the "other" to the promotion of interreligious dialogue and collaboration.[42] This shift occurred during Vatican II with the declaration *Nostra Aetate*, proclaimed by Pope Paul VI on October 28, 1965, which unveiled the new framework for the Church's religious dialogue with other religions. It said:

> Since the spiritual patrimony common to Christians and Jews is thus so great, this sacred synod wants to foster and recommend that mutual understanding and respect which is the fruit, above all, of biblical and theological studies as well as of fraternal dialogues. . . .
>
> Since, in the course of centuries, not a few quarrels and hostilities have arisen between Christians and Moslems, this sacred synod urges all to forget the past and to work sincerely for mutual understanding and to preserve as well as to promote together for the benefit of all mankind social justice and moral welfare, as well as peace and freedom. . . . The Church, therefore, exhorts her sons, that through dialogue and collaboration with the followers of other religions, carried out with prudence and love and in witness to the Christian faith and life, they recognize, preserve and promote the good things, spiritual and moral, as well as the socio-cultural values found among these men.

The Church follows two strategies to preserve its borders. One is dialogue with other faiths to promote respect, understanding, and brotherhood, and the other is admonishing Catholics who have gone astray in their understanding of the requirements of the faith and bringing them back to the fold. For example, the Congregation for the Doctrine of the Faith and the Congregation for the Clergy may intervene to silence a theologian who is developing a biblical interpretation contrary to official Church teaching, excommunicate a bishop who does not follow the directives of the Holy See, or remove or banish such clergy from their posts in an effort to prevent them from challenging the tradition of the Church.[43]

Additionally, the Vatican has developed a global network of informational agents through its nuncios, who are assigned to different countries to inform about problems in local dioceses and the relations between the Church and various states. These apostolic delegates with ambassadorial rank are at the avant-garde of the Vatican's web of diplomats and are best equipped to assess local situations objectively, albeit a bit zealously, and then pass the information on to the Holy See. In the hope that they might be able to resolve a problem at the local level and to protect their flock, bishops, for obvious reasons, are less likely to report everything they know about a priest or their diocese to Rome. Sometimes they do not understand why the Holy See wants to intervene in a matter that they may perceive as less serious. In some cases, a bishop may appear reluctant to collaborate with Rome in dismissing a theologian, for example, in order not to alienate the majority of the clergy of his diocese.

GLOBAL NETWORKS AS MECHANISMS OF ORGANIZATIONAL STABILITY

In its organizational structure, the Catholic Church displays two constitutive operational features: a hierarchy and global network.[44] These features intertwine and complement each other. The vision often projected of a hierarchical Church is misleading because it identifies only one aspect of the equilibrium and ignores the other, which is equally important and central. Hierarchy brings order, administrative stability, and generational continuity to the Church, whereas the network legitimates the central authority, provides room for experimentation, and engenders the reproduction of the system because of the solidarity mechanism that is in place.

The various units of the global network have provided different social contexts in which dissent, different practices, and new religious movements—which may rejuvenate the wider membership—take place. There are several ways in which units have contributed to the rejuvenation of other units and the entire network. Roger Finke and Patricia Wittberg attribute the revival of Catholicism over the centuries to religious orders that have been able to experiment within the larger structure without causing segmentation and fragmentation, as happened in Protestantism.[45] According to this view, network governance is capable of engendering stability because of a number of things it makes possible. It may absorb the positive outcomes of experiments undertaken by any of its nodes. This laissez-faire practice in global Catholicism prevents schism, provides context for dissent, and constitutes a source of revival for all concerned.

The network provides the global church with a social infrastructure for mutual support. When the headquarters is in financial need, it calls on the network by invoking Article 1271 of Canon Law, which declares that "according to the resources of their dioceses, bishops are to assist in procuring those means which the Apostolic See needs . . . to offer service properly to the universal church." Then money from the dioceses is funneled to the center. When a node is in need, the Holy See either makes a donation or coordinates with another diocese to help that node. A node may develop an ongoing effort to help another without waiting for a request from the Vatican. One can think of the "partnerschafts" established by the German Catholics to help poor peasants in the Andes in Peru.[46] A local node may also directly ask another node for help. In addition, programs developed by bishops in the Global North are in place to help dioceses in the Global South. For example, the German bishops have established a special fund—Misereor—to help economic development projects in Latin America and elsewhere. French bishops did the same with the Comité Catholique Contre la Faim et Pour le Developpement, an NGO that they subsidize annually.

Nodes serve as communities of refuge for people from other nodes. Catholic refugees often receive help from the Catholic community in their new place of adoption. Diasporic priests are incardinated into their new dioceses, sometimes to serve offshoot communities from their homelands. Sometimes it is a permanent refuge, and at other times it is transitory, such as when a student travels to complete doctoral or postdoctoral studies and then returns home after graduation.

The global network serves as a support mechanism for doctrinal revival, financial solidarity, the exchange of human resources, and internodal migration. Without this network, the headquarters would be prone to extinction in times of acute crises. The network comes to rescue members and units in crisis and, in the process, sustains itself and ensures its social reproduction.

This global network of relationships may assign specific tasks to certain nodes or may evolve over the years. For example, Switzerland provides the members of the Holy See's police corps (the Swiss Guard); the United States and Germany provide a larger share of the Holy See's budget than do any other countries; Italy provides the employees of the Curia and Vatican City; Western Europe provides the professors who staff the Gregorian University; and until recently, Italy also provided the majority of popes and Curia cardinals.

The network is strengthened by its ability to empower both the headquarters and other nodes. For the most part, problems in the network can be resolved internally. Additionally, the network is strengthened by a set of laws (canon law) that provide precise channels and procedures to follow. The network maintains its vibrancy through this set of laws it develops and implements. In the mid-1990s, when the Vatican faced a deficit in its operating budget, it invoked Article 1271 to remind the bishops of their financial obligations vis-à-vis the Holy See. Not only was the problem resolved, but the status of the budget shifted from deficit to surplus.

The role of the network is again highlighted when the pope addresses his message to a specific church, as the Apostle Paul did through his letters as reported in the New Testament. The message has a global resonance, not only because all the nodes must abide by this new understanding, even though it was not directed at them, but also because it will provide a precedent for later messages and decisions concerning the everyday life of the network.

CORPORATE NETWORK GOVERNANCE

The corporate network governance form requires a certain level of cooperation and sharing in the decision-making process. This does not mean that each node has the same weight in the production of the final outcome, but rather that there is a hierarchy of participation. This participation and networking is calibrated differently depending on what is at stake. Dioceses must inform Rome of any deviation from official norms and practices undertaken in their area of jurisdiction so that corrective palliatives or remedies can be implemented. Often, such practices are made known to Rome through informal channels by parishioners who want to uphold past traditions. If Rome finds out that a diocese has withheld information, it will force it through its nuncio to desist, consent, and submit.

The double vocation of the local church speaks to the globalized nature of the institution. On the one hand, it has all the dimensions of the local place: congregants are citizens of the same nation-state, speak the same language, and partake in the same culture. On the other hand, it also has global linkages and responsibilities as part of a global organization. For the local people, the local church translates the view of the network, and for the network, it translates the view of the local people. This explains the singularity of nodes and the ways in which they contribute to the well-being and sustenance of the network. This also explains why the top-down hierarchical vision advanced by some analysts misses the network aspect of governance that reflects the vitality and deployment of global Catholicism.

Corporate network governance is seen not only in linkages between different sites, but also in the composition of the Curia and make-up of the attendance of Vatican II. Large contingents of bishops and cardinals from Africa, Latin America, and North America were prominent participants. Now, non-Italian archbishops and cardinals from Latin America or Africa head some important dicasters, or departments, of the Curia. The very fact that the pope agreed, at the requests of Cardinal Joseph Frings of Cologne and Cardinal Achille Lienart of Lille, to discard the document prepared by the Curia as discussion points for Vatican II and allowed the council fathers to come up with their own agenda shows the extent of shared authority in the network governance of the Church.[47]

Time is a factor in the process of corporate network governance because it manifests itself in the various phases of the Church's implantation into a community. There is a conquest phase, which needs support from the network to establish a successful enterprise. There is a consolidation phase, in which institutions are built and community life takes on a routine fashion. There is a protection phase, during which authorities might deal with infringement on Church jurisdiction. Protection can be against persecution, expulsion, vandalism, internal dissent, the selling of Church property, or any local scandal. Each case calls for specific intervention by the network tailored to the issue at hand.

What happens at a site may trigger a change of management style that can affect the way in which the network influences the Vatican. Management may emphasize expansion, development, control, harmonization, or doctrine. Managerially speaking, corporate network governance can require different styles of practices and different forms of relations between sites (friendly, amorphous, or tense). This demonstrates how central management is susceptible to the nodes' influence and makes itself flexible to deal with different outcomes while maintaining organizational stability.[48]

Corporate network governance operates in a specific religious context in which the doctrine is universal, the organization is multinational, the actors are mobile and transnational, the decision-making process is shared, and the global institution is stable. The Catholic Church has exerted its corporate leadership by controlling information as it decides what is orthodox and what is not, regulating practices as it decides on behavior that is acceptable or not, and managing the decision-making process as it decides who can be a priest or bishop as well as prevents women and others from accessing the ranks of the clergy hierarchy. No central authority plays that sort of police role in Islam or Judaism.

Due to their Episcopal ordination, the pope and the bishops have equal status theologically speaking, but they differ in terms of ranking order because the pope is *primus inter pares* (first among equals) and because the pope's jurisdiction takes precedence over that of local bishops. In a sense, the decision-making power of the bishops is limited, and, therefore, in this arena they can be seen as "administrative agents of the Holy See."[49]

In the context of corporate governance, bishops are considered managers because they are the administrators of their dioceses, entrepreneurs because they are supposed to develop new areas to serve the expansion of the Church, and spiritual leaders because they are expected to lead the flock in matters of doctrine and rituals. Although corporate network governance assumes diversity in forms of local governance, it maintains homogeneity as the rule, because it calls upon each node to follow the same standards, despite the fact that some nodes are allowed to maintain their different traditions. This happens in the case of Orthodox churches that are in communion with Rome. These are strategies of global control exerted by the corporate headquarters.

Although in the past, the bishop was responsible only for what went on in his diocese, now, because he maintains contacts with diasporic congregations overseas, visits them, sends them pastoral letters, conducts fundraising among them, appoints priests to them, and establishes contacts with the bishops of the foreign dioceses where they are incorporated, the local bishop has acquired a transnational role, in addition to the global role he already plays as a member of a global network of nodes. In that capacity, he participates in synods to contribute to Church governance and doctrine and, in some cases, to the selection of the pope.

Because Vatican City is also a state, the members of the faith who reside there are also theoretically citizens of that state. This is not to say that they routinely claim that citizenship; however, in special circumstances (e.g. war, persecution, natural disaster), it may occur. For example, some people were granted refuge at the Vatican during World War II. Refuge from persecution is sometimes granted to clergy in terms of a new appointment to the Vatican. The Vatican may grant a passport and a visa to a Catholic clergy or lay person who may be conducting a special mission on behalf of the Church and who, because of local laws, cannot get a passport and visa in his own country. This happened when Archbishop John Joseph Cardinal O'Connor traveled to Israel to help solve the Israel-Palestinian conflict. In contrast, Judaism and Islam rely exclusively on the good will of the government where they reside to issue such documents.

Although theoretically it is possible to seek refuge in Vatican City when a Catholic is persecuted, in practice, given the size of the Vatican, this cannot be done on a large scale or even on a small one. Here again, the principle of subsidiarity comes into play. Church policy has been to let the local clergy deal with the departure of the persecuted and the clergy in the receiving country to deal with their resettlement and integration. Here, the Vatican state is different from other states. The spiritual homeland is not available for resettlement, as in the case of Israel, for example. And a Muslim who is persecuted cannot make a claim to resettle in Saudi Arabia, except if he or she is Saudi Arabian by birth. All other Muslims who make such a request will be treated as foreigners on a case-by-case basis with no guarantee of resettlement there.

Because the headquarters of the Catholic Church is not located in the same country as its holy sites in the holy land, their governance and protection is by proxy. This differs from the system of co-governance between the religion and the nation-state that we find in Judaism and Islam, where the headquarters and holy sites are in the same country. In addition, because it is not the original center of the religion, Rome had to impose itself on Jerusalem, the site where the primitive church as an organization or institution was born, and on Avignon, where the headquarters of the Church was once relocated. This problem of the primacy of one site is a global issue because it concerns extraterritorial communities and global leadership. The imposition can be done by force (military, political), by an orthodox interpretation of Scripture, or by tradition. If done by force, it can be undone by force by another group as well. If done through exegesis and tradition, although it could be challenged, it has a better chance to survive. Rome follows the exegetical route through its Petrine Doctrine in reference to apostolic succession and the site of the burial of Peter as the legitimate place for the central administration of the Church.

Corporate network governance is at work even during the interim period that succeeds the death of a pope and precedes the election of a new pope. The conclave for the election of a pope is a global event because the electors—cardinals—come from everywhere on the globe to cast their ballots. The new pope is elected to govern not simply Rome, but the universal church. Upon accepting the selection, the pope gives the "Urbi et Orbi" blessing to the city of Rome and to the world, which symbolically reinforces his global status.

Catholicism, Judaism, and Islam confront the same problem: the site of origin competes with the diaspora and other nodes for prominence, whether because of occupation, migration, or conquest. In Catholic Christianity, the faith has taken root elsewhere—Rome—which, as a new site, has redefined itself as the headquarters whereas the ancestral homeland has become peripheral, at least in matters of the management of the religion. The corporate structure was developed outside the original homeland because the main players relocated elsewhere. From this position, they maintained coequal relations at first, but then these relations became hegemonic.

CONCLUSION

The global deployment of Catholic Christianity has resulted in the use of a form of network governance that reflects both the values propagated by the faith and the imponderables of trans-border geographical expansion. In the process, the faith has benefited from overseas input including the new infrastructural or institutional matrix that molds its mode of engagement in diverse environments. Catholic Christianity has established itself as a global religion and used a corporate network governance structure, which displays singular features different from any other models discussed in this book.

The corporate network governance model has evolved over time to meet the expansive needs of the global Church. On occasions, Canon law establishes the norms for such a linkage, yet at other times local practices are simply formalized, as we have seen in the case of the Eastern Christian churches. In other words, this is not a preplanned model applied everywhere, but rather a model that has emerged based on practice and experience. It both captures the vertical axis and the horizontal connections of the Church that are the centrality of the papacy and the role of bishops in that structure of network governance.

The structure of the network management of the Catholic Church has also evolved over time in response to internal and external crises. One can think of the role of the Holocaust in the establishment of an Interfaith Dialogue Department in the Curia. In addition, the management of the corporate network governance of the institution can also be seen in the evolution of Church practices. For example, major doctrinal, liturgical, and administrative changes are discussed and agreed upon by the network leadership structure during synods before they can be implemented. The initiative for such changes comes either from the Vatican or from any node of the network, oftentimes as a result of local experimentation. In this light, one may add that nodes contribute to both the exercise of the network leadership and the orthodox interpretation of the content of the faith.

Two of the peculiarities of network governance are that it accommodates nodes that operate in diverse environments and that some units are stronger than others. In such an environment, the network gains its stability from navigating in an uneven architecture of relationships. The leadership network may develop differentiated policies to account for the uniqueness of the position of some nodes, as we have seen in the case of the relations of the Holy See with Eastern Catholic Christianity, or it may simply establish differentiated rules of engagement to define modes of interaction among nodes and between nodes and the corporate hierarchical structure of authority.

Corporate network governance is a mobile and variable site of crisscrossing financial, doctrinal, administrative, leadership, institutional, and infrastructural networks. These diverse networks provide the backbone for the governance of various dimensions of the Church. They crisscross because they are in synergy with each other, because no one bureaucracy can handle all of these aspects of governance by itself, because of the interdependence of nodes, and because the success of the global management of the religion depends on the collaboration of each participating node, some with more responsibility than others.

The analysis of the corporate network governance espoused by the Catholic Church reveals the extent to which such an operation is a reflection of the beliefs and moral values the faith propagates. It also reveals that corporate network governance has various layers, each with its own ecology of practice and its unique contribution to the success of the management of the whole Church. Corporate network governance holds the faith together in its multiform institutional arrangements, its multiplex geographical settings, and its make-up of multiple ethnicities and nationalities.

4 Global Islam and Distributed Cellular Network Governance

> Today mosques fall under either government or private sponsorship. In the case of a government mosque, the buildings and staff receive financial support from the state. The Ministry of Religious Affairs or its equivalent hires a professional preacher and monitors his performance. By contrast, a private mosque receives funds from charitable associations. The congregation usually sets the agenda and selects the preacher.
>
> —John L. Esposito, (ed.), The Islamic World: Past and Present

The conquest of the Middle East, North Africa, and part of Europe by Muslim Arabs during the medieval era constitutes a particular form of the overseas expansion of the Islamic faith.[1] Conquest in this case did not simply mean the appropriation of a foreign territory, but also implies the forced conversion of newly created subjects to the religion of the conquerors. However, conquest has not been the only mode of global expansion and enlargement of the membership of the faith. Emigration from the homeland and implantation in a host land, which results in diasporization, is another mechanism of geographical expansion as immigrants develop faith-based communities abroad. This chapter does not study the conquest mode of the globalization of Islam or the once-prominent role of the caliphate in the governance and practice of Islam in Muslim-dominated countries and immigrant enclaves. That imperial form of global religious governance has its own territorial logic and historical context. Rather, the chapter presents an analysis and exegesis of how contemporary Wahhabi Islam deploys and governs its global network of mosques and practitioners and how it manages its relations globally with Muslim communities of different confessional traditions.

WAHHABI ISLAM

Wahhabism is the dominant interpretation and practice of Islam inherited from the teaching and preaching of Muhammad Ibn Abd al-Wahhab (1702–1791).[2] Seen from the standpoint of the believers, it is a restorationist, ultraorthodox, or ultraconservative form of Islam, because the religion is supposed to be a return to a purer, more authentic, orthodox, and austere form of Islam maintained by a small group of people prior to the independence of Saudi

Arabia and whose descendants now govern the country. However, because in the West people tend to associate this form of Islam with fanaticism, extremism, terrorism, and intolerance, the practitioners in Saudi Arabia do not refer to their religion as "Wahhabi Islam."[3] Instead, they call themselves simply "Muslims" or "Monotheists" and consider the term "Wahhabi" to be a form of Western bias toward the people who practice the faith. This view was well expressed as early as May 11, 1929, by King Abdul Aziz al Sa'ud in a speech delivered at the Royal Palace in Mecca:

> They call us the "Wahhabis" and they call our creed a "Wahhabi" one as if it were a special one . . . and this is an extremely erroneous allegation that has arisen from the false propaganda launched by those who had ill feelings as well as ill intentions towards the movement. We are not proclaiming a new creed or a new dogma. Muhammad Ibn Abdul Wahhab did not come with anything new. Our creed is the creed of those good people who preceded us and which came in the Book of God (the Quran) as well as that of his Messenger.[4]

The predominance of Wahhabism in Saudi Arabia has a genealogy characteristic of rejected knowledge. From the perspective of practitioners, rejected knowledge is knowledge that was once dominant, but became peripheral or subaltern over time because of other truth claims by people in power. According to these practitioners, with the seizing of the nation's leadership and the help of oil resources, this rejected knowledge can now make a spectacular comeback and regain its lost center-stage position. In this scenario, Wahhabism has evolved from center to periphery and back to the center. This interpretation is a better explanation of the genealogy or itinerary of Wahhabi Islam than resorting to abusive discourses that label its practitioners as extremists.

This chapter captures this mobility in its new, ascendant phase. Although the Saudis consider their Wahhabism to be orthodox Islam, they see other forms as deviant, employing post-Muhammad ritual practices that have been added by outsiders to the faith. Regardless, this conservative interpretation of Islam cannot be used as a panacea to justify the Saudi government's shortcomings in areas of democratic practices, human rights, and intolerance of other religious systems.

A distinct aspect of Wahhabism is the priority it gives the Hanbali school of jurisprudence over the other three schools, the Shafii, Hanafi, and Maliki, which offer different interpretations of Islamic law. Since 1928, Hanbali books, which provide the legalistic interpretation that serves as the backbone of the practice of Wahhabi Islam, have been the main pillars of the Saudi government's legal system. Local judges use them to arrive at their court's judgments, although other schools of legal thought are consulted for certain cases, even if they are not the dominant interpretation.[5]

For our purposes, however, the most important characteristic of Wahhabi Islam is the distributed cellular network form of governance that it promotes, which seeks the interpenetration of the secular state and the faith. Instead of there being a secular government with a constitution, on the one hand, and the religious group with its holy scripture on the other, one is the face of the other. That is, the state's constitution is the holy scripture, and the legal interpreters are religious scholars with academic expertise in Islamic jurisprudence.

SAUDI ARABIA AS THE HEADQUARTERS OF WAHHABISM

Like Judaism and Catholicism, Wahhabi Islam has a country, Saudi Arabia, that it refers to as the site of its headquarters.[6] Saudi Arabia, which officially reassembled itself as an independent and sovereign territorial state in 1932 under a self-appointed king and was duly recognized by international law, is not only the birthplace of Wahhabi Islam, but also the location of the two most holy sites of Islam, Mecca and Medina. Any study of Wahhabi Islam must pay attention to its development in Saudi Arabia. However, because this chapter analyzes the management of global governance, an exclusive focus on Saudi Arabia is not necessary, because one can penetrate the transnational network from any of its distributed or dispersed nodes. In so doing one will be able to unveil its transborder interconnections and translocal modes of practice that account for the governance and vitality of the global network of sites and practitioners.

The cohabitation of the "religious" (the Wahhabis) and the "political" (the House of Saud) in Saudi Arabia has led to the production of a governmental structure shaped by its Wahhabi doctrinal orientation with a global leadership role made possible by the annual pilgrimage (the hajj) to Mecca and the rich oil fields that Saudi Arabia controls. Antoine Basbous speaks of the synergy in Saudi Arabia between the civil and religious powers, which he identifies as "parallel dynasties," to explain how the king and government officials maintain their hegemony over the state apparatus in a long-lasting strategic alliance with the religious Wahhabi group that legitimates their secular authorities. The High Committee of Senior Ulama, which constitutes the nerve center of this dynastic religious system,[7] provides religious *encadrement* to the Saudi state at home. Abroad, it exerts influence through the Islamic Affairs Office of the Saudi Embassy (whether in Paris, Washington, London, Istanbul, Islamabad, or Tokyo), which it partly staffs. An ulama, or legal scholar, oversees that section in each country where a Saudi embassy is established and provides theological guidance to the local brethren and also to official government staff employed in these diplomatic outposts. The ulama group of religious leaders constitutes an important link of the kingdom to the overseas tentacles of Wahhabi Islam and is a coordinating mechanism with an advisory role vis-à-vis the Saudi government.[8]

Saudi Arabia–based Wahhabi Islam sees itself as a religious movement or religious formation with real and potential tentacles all over the world that are either its extensions (diasporic enclaves and the overseas mosques that it finances), competitors in the faith (other Muslim entities that need persuasion to follow the right path, meaning Wahhabism), or simply infidels who have not yet experienced enlightenment by the holy book of Koran and, therefore, need Islamic conversion. Global governance in this perspective implies strengthening or reinforcing existing or dormant linkages among the sites and brethren, maintaining a fruitful dialogue with other Muslim communities (Muslims of the Shiite tradition and liberal Muslims of the Sunni tradition), developing friendly relations with the other Abrahamic faiths (Judaism and Christianity), helping members in need (whether from natural disasters, chronic poverty, as refugees, or in war zones), governing the borders of these relations, protecting Muslim minorities worldwide, and influencing the policies of other states and international organizations that might be detrimental to Muslims.[9]

Global networks have plural nodes with different orientations and different strategies of managerial practice, and this multiplicity is seen in the case of the extraterritorial tentacles that relate differently to Saudi Wahhabism, whether they are located in Muslim countries, Western industrial countries, or developing countries of the Global South. These differences exist even within relations with other Muslim countries in reference to specific foreign policy or national security issues, because they are differently positioned vis-à-vis Saudi Wahhabism. For example, from the standpoint of Saudi Arabia, Saddam Hussein's Iraq was a hostile neighbor during the invasion of Kuwait. Iran is a competitor partly because of its large Shiite population. Jordan is a supportive neighbor, but Egypt is not a dependable ally. Syria is a strategic partner in regard to the Palestinian-Israel conflict, and Kuwait is a collaborator and reliable friend. Saudi Arabia has developed different international strategies to deal with so-called infidel countries because of the variable size of Muslim populations in their midst. For example, different strategies may arise to deal with nations where there is a history of friendly relations (e.g., the United States and England), a history of inimical relations (e.g., the now-defunct Soviet Union), or no official diplomatic relations at all (e.g., Israel).

The peculiarity of Wahhabism is that it is the exclusive religious doctrine and official religion of the Kingdom of Saudi Arabia, whose practice is mandated by law in general and enforced by the state religious police (Muttawa), in particular.[10] The Koran, serving as the constitution, and shariah law, acting as the official jurisprudence system of the country, guide the actions of government, the everyday life of citizens, and the institutions of society. The petroleum resources of the kingdom allow it to spread its Wahhabi doctrine across the globe, a sustained effort that has no equivalent in scale in other Muslim states.

The strictness of Wahhabism and the everyday policing of its boundaries contribute to the successful reproduction of its traditional practices. Wahhabism appeals to the myth or reality of the simplicity and purity of the original primitive community, which bears the orthodox tradition. In the view of the Saudi Wahhabis, their native culture, history, and geography make them more qualified than converts from abroad to unveil, preserve, and develop the orthodox tradition through interpretation and ritual practices. The effort of the Wahhabis to control the dominant ground and the key to the orthodox interpretation of Islam is predicated on the religion's birthplace and the location of its most important religious site, Mecca. This parallels the struggle that the Church of Jerusalem—and with it the Eastern Orthodox Church—fought when the Gentile Church of Rome established itself as the faith's administrative headquarters. The Church of Jerusalem, despite being the first church in the land where Jesus was born, preached, died, and resurrected, became subject to the hegemonic diasporic Church of Rome. The Wahhabis of Saudi Arabia do not want this to happen to them. As descendants of the primitive Meccan religious community, they claim to represent the orthodox practice of Islam and claim to be entrusted by reason of tradition, location, and practice to serve as the privileged custodian of the faith and the holy sites.

Among the three monotheistic faiths, only Wahhabi Islam and Judaism can pretend to be the direct custodians of holy sites in the homeland on behalf of the faithful because of their ability to provide protective security and prevent vandalism or desecration. In both cases, the government (Saudi Arabia; Israel and the Palestinian Authority) maintains protection over the sites. As we noted, because the Catholic Church headquarters is located in Rome, it depends on the government of Israel and the Palestinian Authority, for the most part, to protect its sites in the holy lands. The Saudis are in the same predicament vis-à-vis the protection of the third holiest site of Islam, located in Jerusalem, where it is believed the Prophet Muhammad ascended to heaven.

Thus, by appealing to the origins of the faith, by virtue of residing within the country where two of the three Islamic holy sites are located and serving as the custodian of the sites, by hosting pilgrim guests from all over the world, and by financially supporting Islamic centers abroad, Wahhabi Muslims in Saudi Arabia have played a global leadership role.

DISTRIBUTED CELLULAR NETWORK GOVERNANCE

The distributed cellular network form that characterizes the unfolding of the management of global Islam is different from the previous cases of network governance studied in this book.[11] It is a network in which the nodes are cells that are autonomous, but that also relate to and rely on each other in a fluid manner. For example, like Judaism, Islam has a latent global system of security, but unlike the governance of the security apparatus in Judaism, its operation is not centralized, nor is it interactive. Security is left to the local nation-states and is mostly directed to the enforcement of mores (rules of behavior) and to ensure good behavior in terms of dress code, attention to prayer times when shops are supposed to be closed, and male-female relations in the public sphere. The religious police in Islamic countries serve as both intelligence and police. The Muttawa of Saudi Arabia, known as the Commission for the Protection of Virtue and Prevention of Vice, comes to mind. Each Islamic country has its own way of pursuing this issue, some more repressive than others. There is no mechanism that links these diverse local systems through which they could engage to protect each other globally. This is a fragmented system that has an exclusively national/local vocation. A behavior that is restricted in one country may be allowed in another.

In the business-and-management literature, global cellular organizations are seen as a by-product of the knowledge age that reflects the independence of a firm that is still connected to other firms in a way that provides mutual sustenance. In the cellular model, firms function like cells but enjoy a "combination of independence and interdependence," features that allow them to behave in different ways at different times. Thus, "like a cell in a large organism, each firm has its own purpose and ability to function independently, but it shares common features and purposes with all of its sister firms."[12] This model, with some modifications, describes the global governance model of Wahhabi Islam.

Distributed cellular network governance is a form of dispersed leadership that involves every believer in the faith's maintenance and management, expansion, and everyday operation. This form of governmentality is appropriate because Islam does not have a formal clergy or sacramental rituals for which such a clergy could be responsible. Thus, a third party does not mediate the individual's relationship with God. Instead, everyone is called on to contribute to this form of governance according to each person's physical and intellectual capacity, resources, talent, and status. For example, some help lead the weekly prayer at a mosque, whereas others help interpret the Koran, and still others contribute financially or assist members of the government in applying shariah law.

Distributed cellular network governance takes different forms. It is carried out through the mutual dependence of nodes in order to influence the global network, the distribution of financial aid, the establishment of Islamic centers and schools, and overseas training. It is also carried out through the protection of other Islamic communities and through the protection of the doctrine and rituals from corruption and unorthodox or liberal interpretation. The effective deployment of distributed cellular network governance is particularly seen in the areas of global proselytism, the mobility of religious leaders, the protection of the borders of the faith, and the use of telecommunications to reach the nodes in the global network.

GLOBAL PROSELYTISM

Natana J. DeLong-Bas notes that the Saudis engage in "aggressive proselytizing" to increase the number of converts, but not in holy wars, military invasions, or conquests.[13] They do so through charity and missionary work abroad, using petrodollars to finance their religious projects.

In carrying out the task of proselytizing converts, Wahhabism is circulated around the globe by specialized institutions sponsored by the Saudi state, such as the World Assembly of Muslim Youth, the Muslim World League, and other government agencies. Distributing free copies of the Koran at home and abroad is a proselytizing practice of the kingdom. Sometimes the kingdom targets a specific country for this form of proselytism, as when King Fahd sent free books to Muslims in London. At other times, the kingdom seeks to reach many nations simultaneously. In 2004, close to two million copies of the holy Koran, translated into Urdu, French, Filipino, Indonesian, and English, were distributed inside the kingdom to local and foreign pilgrims at the end of the hajj or sent abroad for distribution to foreigners through Saudi embassies and consulate offices, Islamic cultural centers, and mosques.[14] During the last decade of the twentieth century, the kingdom published approximately ten million copies of the Koran annually. Not all of these books are in Arabic—some are in other languages such as Chinese, English, or Turkish.[15] What is unusual here is the use of state institutions for such a practice.

Several institutions and individuals in Saudi Arabia are involved in religious projects overseas. For example, in 2005 the World Assembly of Muslim Youth (WAMY), headquartered in Riyadh, undertook and began construction of four mosques in Yemen.[16] In 2003, WAMY built three mosques in southern Asia and a new mosque in Sudan.[17] Ministers of the Saudi government and members of the royal family have donated their own money to the construction of mosques abroad. For example, Prince Sultan, who in 2004 served as the second deputy premier in the Ministry of Defense and Aviation and as inspector general, was the main financial contributor to the construction of a mosque in Kazakhstan.[18] In 2002, the Jeddah office of WAMY boasted that it already had financed "206 Qur'an memorization circles in eight countries, and [had] distributed over 150,000 educational and Islamic books worldwide."[19] The Saudi government, through its local embassy or an organization sponsored by the regime, such as WAMY, finances *madrassas* (Islamic schools) in a few Islamic countries.[20] These schools exist and succeed because of the financial aid they receive from the kingdom. The Muslim World League, headquartered in Mecca, also dedicates itself to missionary work in order to spread Islam.

The government of the Kingdom of Saudi Arabia has devoted an entire ministry to spreading Wahhabi Islam outside its borders. The Ministry of Religious Affairs, Endowment, Da'wa, and Religious Guidance, as it is known, states the following among its principal goals:

> Bringing out the Kingdom's efforts in the field of Islamic mission and reinforcing the Muslims abroad. . . . The call to God outside of the kingdom in both Muslim and non-Muslim territories. Mission to elevate the intellectual and scientific levels of Muslims abroad and reinforcing their Islamic identity. Looking after the Muslim minorities and immigrants in the world and aiding them in protecting their rights to practice their religion, among other rights. Strengthening Islamic institutes, foundations, and academies abroad and preserving their affairs and promoting their goals and activities.[21]

To enhance the spread of Wahhabi Islam, the kingdom has been engaged in sponsoring and financing a global network of mosques, Islamic centers, and schools. In 1996, the website of the Saudi embassy in Washington, D.C., noted that "the Kingdom has set up, or contributed to the building of, a total of 210 Islamic Centers and 1359 mosques worldwide, in addition to donating generously to another 1569 such projects."[22] These institutions, which received generous Saudi financial support, are found in Geneva, Madrid, London, Bonn, Brussels, Rome, Zagreb, Lisbon, Vienna, Edinburgh, and Moscow. In the United States, they can be found in Maryland, Virginia, Ohio, San Francisco, Los Angeles, and Washington, D.C. They can also be found in Canada, Latin America, Australia, Japan, and other Pacific Rim countries. By 2000, the Saudis had also financed Islamic centers and mosques in Fairfax, Virginia; Fresno, California; Denver, Colorado; Harrison, New York; Chicago, Illinois; San Francisco, California; Toledo, Ohio; Boulder, Colorado; New Brunswick, New Jersey; East Lansing, Michigan; Raleigh, North Carolina; and Tucson, Arizona; as well as in Fiji, Indonesia, Mauritania, Djibouti, the Maldives, Cameroon, Somalia, Gabon, Burkina Faso, Tanzania, Senegal, Nigeria, Sudan, and Algeria; Toronto, Ottawa, Calgary, and Quebec, Canada; Buenos Aires and Rio de Janeiro; Victoria, Australia, and Auckland, New Zealand.[23]

Establishing, sponsoring, and financing schools abroad is done for three purposes: to provide educational facilities to Saudi students who live abroad (children of diplomats or immigrants); to indoctrinate non-Wahhabi Muslim and non-Muslim students with the Saudi's version of Islam; and to facilitate the spread of the Arabic language.[24]

The financial dependence of some foreign mosques on Saudi money for their functioning may explain the government's control over the appointment of a coterie of imams—leaders of worship—outside the kingdom. In 1995, on the inauguration day of the Islamic Center in Cincinnati, Ohio, Prince Bandar Bin Sultan Bin Abdul Aziz, the Saudi ambassador to the United

States, declared that "the Saudi Embassy in Washington DC will appoint an imam . . . and allocate an annual budget for the new center, which contains a mosque, a cultural center, and a school for teaching Islamic sciences and the Arabic language."[25] The embassy, in this case, functions as a coordinating mechanism in the expansion, deployment, and functioning of satellite institutions located outside the territorial jurisdiction of the kingdom.

The government has either created or financially supported a number of international organizations with global outreach to disseminate Islam and revive links with minority Muslim communities throughout the globe. In addition to the Muslim World League and WAMY, these organizations include agencies such as the Higher Committee for the Collection of Donations for Muslims of Bosnia-Herzegovina and the International Islamic Relief Organization.[26] In the same vein, one of the mechanisms used to encourage the study of the Koran worldwide, to elevate the status of scholars who subscribe to the Wahhabi interpretation of Islam, and to propagate the faith is the King Faysal Prize, awarded to Islamist scholars or converts such as the French philosopher Roger Garaudy, the Pakistani Abou Ala'a Al-Maoudoudi, and others of Indian, Nigerian, Egyptian, or Afghan origin.[27]

The Jeddah-based headquarters of the International Islamic Relief Organization (IIRO) is a central mechanism in the spread and sustenance of Wahhabism outside the kingdom. All proselytizing religions incorporate humanitarian projects in their missionary work, which includes running hospitals, orphanages, and schools. Wahhabi Islam is no different. The IIRO functions both as a relief agency effort and a proselytizing channel. In 1995, it ran "108 schools, 14 colleges and 17 vocational training institutes in 69 countries." At that time, it also sponsored "2,100 students and 559 teachers."[28]

An important element in the global strategy of Wahhabi Islam is the influence exerted by the Saudis over some overseas mosques. Sending an imam to provide leadership to a faithful community abroad, paying for the facilities and salaries, or retraining the local imam to become more orthodox—that is, to follow and teach the Wahhabi path—are methods used to influence the religious orientation of foreign mosques. Saudi Arabia has employed all of these mechanisms to spread Wahhabism.

Each strategy is deployed to confront specific issues. Sending a missionary to a local mosque requires surmounting potential language problems and understanding and appreciating local traditions to prevent conflict and foster harmony. Financial support requires constructing or activating a social infrastructure, which accounts for expenses and projects. Training young people in Mecca for ministry at home or abroad requires selection, entry, and exit strategies. Therefore, global governance of a distributed cellular network requires a series of governing acts that have their own individualities, because they pertain to the prevention or resolution of specific sets of problems.

Although in the theological logic of Wahhabi Islam, the initiative for global proselytism comes from every member of the *ummah*—the community of the faithful—and, therefore, is not confined to any group in particular, some individuals or network nodes are strategically better located to undertake it than others. Some people use their position in secular society to influence the network in one way or another. Power and wealth generated in civil society place some in better positions to influence the network more visibly than others. In this rotation of leadership, individuals undertake actions on behalf of the faith, not because of a consensus over a policy outcome or because of their own authority, but in the name of Allah. Legitimacy does not come from any human sanctioning authority, but from the execution of the will of Allah.

The extraterritorial expansion of the faith may lead not only to an increase in Wahhabi membership worldwide, but also to the dominance of its interpretation and the peripheralization of other religious traditions. It is competition through expansion. Because other Saudi groups, such as the Shiite minority and liberal Sunnis, do not have the official support of the Kingdom of Saudi Arabia, any missionary work they carry out must be done using their own resources. The Wahhabi group is necessarily in a privileged and more powerful position.

Proselytism requires the existence of people to do outreach, a financial basis to embark upon such a project because of the expense involved, and the goods, such as books and religious paraphernalia, that are needed to make it a success. The Islamic University of Medina, through its Deanship of Service to Society, "distributes and prints material in schools and institutions inside and outside the Kingdom" for this very purpose.[29]

The impact of Wahhabi proselytism can be seen in its activities in Morocco. A Moroccan observer notes the way in which Wahhabi penetration undermines the community's local Islamic practices and its long-held Maliki school of jurisprudence. He points to the books, videos, and audiocassettes in circulation in Casablanca and sees Wahhabi influence not only in the area of religious thought, but also in the imposition of a dress code and the promotion of Saudi Arabian perfumes now used by the faithful.[30]

The global strategy of proselytism in this distributed cellular network follows different organizational channels. Some are totally controlled by the government, and others are autonomous institutions to which the kingdom contributes—for example, the activities of WAMY.[31] Two Saudi Arabian religious entities, the Council for Islamic Call and Guidance and the Supreme Council of Islamic Affairs, oversee this global apostolate or outreach.

RELIGIOUS CONTROL OF THE GLOBAL MOBILITY OF MUSLIMS

Religious tourism or pilgrimage poses many problems to the network governance of global religions. Regulating the global mobility of the faithful requires the collaboration of the state and the religious institutions. This is particularly the case with the governance of the distributed cellular network of Islam. Because Muslims are mandated to go to Mecca at least once in their lifetime, the state must provide the physical and social infrastructure needed to welcome them and must cooperate with both religious institutions in other countries and the governments of other states for the success of such a permanent venture. In this specific area, it is difficult to see where the state responsibility ends and that of the religion begins.

In studying the distributed cellular network governance of the global mobility of Muslims, I will focus on the hajj,[32] including the local organizations that require coordination with the Saudi authorities; extra medical foreign personnel; the quota system, which reflects the total number of pilgrims Saudi Arabia is willing to let in; the practice of barring certain groups or individuals from entering the country; the mobility of students recruited for theological training at the Islamic University; and the mobility of educators and religious leaders in missions abroad. The global governance of this distributed cellular network in each of these aspects requires the application of proper rules and regulations under shariah law. The hajj is one occasion when the global network of nodes is activated, when the central node expresses its global leadership role, when each node is called to adjust itself to the orthodox teaching of the faith, and when a local tradition of interpretation is reaffirmed for the consumption of the global network.

Through the annual quota system, the hajj minister regulates the movement of pilgrims from their countries of origin to the holy sites of Mecca and Medina. This border policing also applies to the minor pilgrimage called the umrah, a series of ritual acts symbolic of the lives of the prophet Abraham and his wife Hagar and of solidarity with Muslims worldwide. The quota system that controls migration flows to Saudi Arabia for the purpose of performing these two pilgrimages was established in 1987 on the occasion of a meeting in Amman, Jordan, of the ministers of foreign affairs of countries with sizable Muslim populations. The method used since then to assign quotas to countries is that "the number of pilgrims from each Muslim country is determined by dividing the total population of that country by 1,000."[33] Prior to 2001, up to 1.5 million were allowed in, and thereafter, there was a projected annual adjustment for 10 million per year. While visiting the Saudi Embassy in London in June 2001, the hajj minister spoke to a group of British Muslims and informed them that the new regulations will give "umrah performers more time to enjoy visiting Makkah, Madinah, and other historic places."[34]

There are several conduits through which information is imparted by the Saudi government to prospective pilgrims. The embassy is a key intermediary that distributes information about the pilgrimage to various mosques and Islamic centers in other countries. Sometimes the minister of pilgrimage visits certain countries, meets with Islamic representatives, and explains the regulations for the year.

Although every male adult is expected to make a pilgrimage to Mecca once in his lifetime, the timing of such an undertaking may interfere with the kingdom's border policies, which might reject pilgrims in a given year either because of the quota system, for political reasons, or for health or security concerns. For example, because of the Ebola outbreak, no hajj visas were issued to potential pilgrims from the Republic of Congo in 2004.[35]

The distributed cellular network governance of corporeal mobility in relation to the hajj must be seen in terms of the governance of three vectors: governance of pilgrim mobility from their homeland to Saudi Arabia, governance of their mobility inside the kingdom en route to Mecca and Medina, and governance of the global distribution, to underdeveloped and countries in need, of animals slaughtered for the hajjis. This triple aspect of global governance, choreographed by several government ministries and agencies, concerns external relations, national security, public health, transportation, and religious affairs.

An elaborate system is put in place to control the movement of pilgrims from their homelands to Saudi Arabia. This system is administered by travel agencies known as "hajj tours" in some Muslim countries (in India, for example, there are three well-known companies of this kind: All India Haj Umrah Tour Organisers Association, All India Haj and Umrah Tour Operators Association, and the Karnataka State Haj Organiser's Association) and by the government of the homeland. This system of control is based on the requirements of the Saudi government and addresses issues regarding short-stay visas. Specific issues of concern include eligibility to participate in a given year, security clearances, health conditions, the airlines used for transportation, the site of arrival and departure, lodging and ground transportation, and so on. The inability to meet any of these requirements will lead the government to not grant the visa and to the cancellation of the projected trip. This global coordination is necessary because pilgrims arrive by air, sea, and land and must be accommodated once they land in the kingdom. Sometimes individuals have been denied permission to make the pilgrimage not because they were in violation of the regulations, but because either the kingdom could not accommodate them for that year or the quota assigned to their country had been filled. When this happens, these individuals are given priority the following year.

The ministries of religious affairs in Islamic states are responsible for the management of hajjis in their own countries to ensure a smooth process, to coordinate these activities with their consulates in Jeddah and the Saudi authorities, and to resolve any conflicts or complaints about the Saudi government that emanate from their pilgrims. The elaborate requirements from the kingdom force the homeland governments to invest both time and money in the local management of the hajj.

Ground transportation inside the kingdom has also been the object of an elaborate regime to ensure safety, prevent exploitation by local guides, and ensure order in the deployment of diverse pilgrim contingents (formerly in caravans). Both the Ministry of Transportation and private companies are involved in this lucrative process. In the past, local bus and taxi drivers charged exorbitant fees in an effort to exploit the pilgrims, but now the government sets fare schedules and has established an office to file complaints and for conflict resolution. Disputes can be negotiated and resolved at this level, whereas in previous years, intervention of the homeland government was needed to resolve such issues.

Not all pilgrims use their personal money to pay for the trip. In fact, not all Muslims have the means to do so. In 1997, to allow impoverished individuals to accomplish this act of piety once in their lifetimes, King Fahd Bin Abdul Aziz developed a program in which he sponsored about one thousand pilgrims each year to visit the kingdom free of charge. In 2000, these pilgrims were from China, Japan, Korea, Thailand, Taiwan, Cambodia, Vietnam, Laos, Burma, Indonesia, Philippines, Singapore, Malaysia and Ukraine, whereas in 2003, contingents from South Africa, Senegal, Ghana, Albania, Poland, Japan, and Latin America benefited from his largesse and made the hajj.[36]

Many of these pilgrims arrive in Saudi Arabia as part of a group, such as the Official British Hajj Delegation. This reflects the screening and arrangements made with the Saudis prior to their departure. Such delegations include not only pilgrims, but also preachers who guide them in the performance of rituals and medical doctors who help them in case of illness or physical discomfort. Because of the hajj visa process, both the homeland and Saudi governments know the identity of these professionals. The subdivision of pilgrims by country delegation is a mechanism used to ensure that these pilgrims will return home after the hajj.

The transportation of pilgrims from their country of residence to Saudi Arabia is an object of tough negotiations because huge revenues are involved. These global negotiations ensure not only that Saudi Arabian Airlines (SAA) receives its fair share of pilgrims for the year, but also that the kingdom will know ahead of time the cities from which pilgrims will depart to ensure orderly departures, arrivals, and protection of national security, to screen out potential troublemakers, and to maximize revenues. It was under the terms of such an agreement, for example, that in 2005, the Iranian government (the Iranian Hajj Organization) allowed SAA to transport only one hundred thousand Iranian pilgrims for the hajj of that year, while it gave the airlines permission to accommodate a quarter of a million for umrah.[37]

The number of pilgrims who apply to undertake the hajj could easily overburden the capacity of the embassies if special efforts, such as the development of a special section of the embassy or consulates in charge of hajj and umrah visas, were not planned ahead of time. For example, for the umrah of 2004, the Saudi government issued 697,969 visas to Egyptian pilgrims, 328,785 visas to Iranian pilgrims, 304,722 visas to Pakistani pilgrims, 184,458 visas to Jordanian pilgrims, and 159,606 visas to Syrian pilgrims.[38]

The global governance of Islam is fundamentally a gendered enterprise in which different rules for men and women are developed and enforced and additional requirements for women pilgrims are not applied to men. This is seen in the issuance of visas, in which a specific regulation is added to control the mobility of female bodies. One of the requirements in the *Guide to the Hajj* that the Saudi consulate offices and embassies implement is that "all ladies are required to travel for Hajj with a Mahram [an unmarriageable kin with whom sexual intercourse would be considered incestuous, a punishable taboo]. Proof of kinship must be submitted with the application form. Any lady over the age of forty-five (45) may travel without a Mahram with an organized group, provided she submits a letter of no objection from her husband, son, or brother authorizing her to travel for Hajj with the named group. This letter should be notarized."[39]

Despite all the regulations developed by the kingdom to ensure that the pilgrimage deploys smoothly, there are always groups that remain difficult to control because of their unique itinerary toward Mecca. For example, David Long notes that "many West African Hajjis literally walk across the continent, stopping for as long as a year or so to work and raise money to pay expenses. Many end up at Khartoum, Sudan, with no visas to the Kingdom or travel documents. They not only swamp the Saudi embassy, but their own embassies as well."[40]

The global mobility of Muslim bodies pertains not only to the movement of pilgrims to the kingdom, but also to the movement of Saudi preachers and teachers to foreign countries to carry out religious missions. Wahhabi preachers have been dispatched by the kingdom to such places as Venezuela, Bangladesh, Pakistan, Albania, Brazil, and Argentina to open new vistas for the faith or to strengthen it among immigrants and converts. Preachers are sent to Islamic countries to prevent the more liberal Sunni and Shiite interpretations and mainstream practices from contaminating Islamic doctrine and to non-Islamic countries for purpose of conversion and consolidation. Global governance implies different strategies to meet these different local conditions.

The global governance of the distribution of leftover meat slaughtered during the hajj to feed pilgrims also requires contacts with foreign governments for the selection of countries, to assess the needs of the population, to inquire about local mechanisms of distribution, and to ponder the availability of ground transportation. In 2000, the Islamic Development Bank (IDB) in Jeddah oversaw the global distribution process. The meat was flown from Saudi Arabia through King Abdul Aziz Airport in Jeddah to twenty-seven countries.[41]

Beyond the need to distribute these ritually slaughtered animals, as in the case of kosher foods in Judaism, the production, distribution, and consumption of halal foods—foods permissible under Islamic law—has global ramifications.[42] Halal does not require ritual procedures or the supervision of an imam, but the distribution and commercialization of halal foods has become global.

TRAINING FOR LOCAL AND GLOBAL LEADERSHIP

Like any global institution or firm, Wahhabi Islam trains its religious leaders for global leadership. It does so in many ways and at various levels. One central mechanism used for that purpose is training new leaders in theology and law so that they can develop an orthodox understanding of the holy scripture and the jurisprudence system that guides the incarnation of religious principles in daily life.

As a way of preparing the global leadership of Islam, for many years Saudi Arabia has brought young scholars to the kingdom to study at the Islamic University of Medina, founded in 1961. It is an institution whose ultimate goal is "to impart religious education to students who come from different parts of the world," hoping that in so doing it will "convey the message of Islam to the world at large." The university clearly states the global role that it envisions for graduates, both the Saudis and the international students, and what is expected of them as part of their specialized service to society. They are qualified "to serve on Islamic delegations or as teachers and lecturers for teaching Arabic and Islamic culture in Muslim countries around the world. Graduates are also qualified to lead prayers inside and outside the kingdom, and hold seminars and lectures." The university also states the kingdom's obligations toward foreign students in terms of financial contributions to their transportation, board and room, and living expenses in general. For example, "for outstanding students from overseas, the University offers a wide range of generous benefits, including travel between their home countries and the kingdom, and while studying, free accommodation, textbooks and transportation, in addition to the regular monthly stipend that all students in the kingdom receive."[43] In 2000, these students came from 105 foreign countries.

Global governance in the school context implies not only the building of these structures with Saudi money and the dependence of these institutions on the Wahhabi source of knowledge and interpretation, but also the dispatch of Saudi teachers to ensure the orthodoxy of what is taught. In this context, one understands why the Education Ministry financed the salaries and expenses of forty-six Saudi teachers who were sent in 2000 to Bahrain, Niger, Kenya, the Comoros, Senegal, Indonesia, and the United States and to Saudi academies and schools in Washington, D.C., Moscow, Islamabad, Karachi, Djibouti, and Rabat.[44]

The ultimate goal of this training at the Islamic University is to prepare students for religious leadership in their own communities. This is a key mechanism for the global reproduction and domination of Wahhabi Islam. Such scholars are trained to become imams, spread and consolidate

Wahhabi Islam in their country, maintain contacts with Wahhabi leaders or teachers in the kingdom, and serve as representatives of Wahhabi Islam in their country of residence. As a result of their education in Saudi Arabia, they, more than any others, are familiar with the Islamic theology taught at the Islamic University and are the principal conduits through which Saudi money can be filtered to build Islamic centers and mosques. They are most strategically positioned to be selected as imams by the Religious Affairs Office of the Saudi embassy in their country of residence.

The government of Saudi Arabia annually awards approximately five thousand scholarships to foreign students. Some of the students brought to the Islamic University are not simply there for training, but sometimes for retraining. To ensure that they propagate an orthodox form of Islam and to prevent them from tinkering with the faith, some African American Muslims in leadership positions are also given scholarships for retraining.

IT AND WAHHABI ISLAM

The role of digitization in the efficient management of distributed cellular network governance is not peripheral to the process, but has been a prime mode of communication used to reach believers beyond the boundaries of the Kingdom of Saudi Arabia and across the globe. Religious authorities in Saudi Arabia have used information technology (radio, television, and the Internet) to allow easy access to religious texts through their websites, to disseminate sermons to overseas brethren, and to reach Muslims in other countries.

This practice of producing programs in the homeland and broadcasting them in the host land is a universal phenomenon in most immigrant communities. In the case of Wahhabi Islam, this practice is used because religious activities in the holy sites cannot be duplicated abroad with the same emotional intensity. Television and radio programs will no doubt continue to feed local needs and serve effectively as a means to propagate national folklore, to consolidate acts of piety, and to enhance transnational indoctrination.

Information technology was also used in many ways by the government to facilitate the smooth deployment of pilgrims during the 2005 hajj and by individuals to remain in contact with their families at home. According to Afzal Khan, "The 129 gates to Mecca's Grand Mosque that enshrines the Ka'ba . . . have been fitted with traffic-style lights that turn red when the mosque is full and green when there is space to enter. Up to 700 television sets and five giant electronic screens have been set up in the tent city in Mina to inform pilgrims about traffic conditions. . . . The TV sets also broadcast the holy rituals in five languages."[45]

The kingdom uses a radio service system to reach its foreign audience. The goal of this mission is to propagate the faith and to rectify unorthodox religious doctrines and interpretations of Islam. It broadcasts daily programs either from Riyadh or Jeddah in French, English, Turkish, Persian, Urdu, Swahili, Indonesian, and, of course, Arabic.[46] The Saudi Television Service has been used for the same purpose and has been able to do so via satellite transmission.[47] It is used to project a positive image of the kingdom and of Wahhabism and to share the kingdom's religious activities with the rest of the Muslim world.

Saudi Arabia uses its technological communications infrastructure to broadcast its religious programs to Muslims all over the world, especially on the occasion of religious celebrations. Arabsat (Arab satellite), which has been in orbit since 1985, is used to transmit televised religious programs to other Muslim countries.[48] Television provides a service to Muslims in need of spiritual nourishment, and it also reinforces the global leadership role of the kingdom in religious matters.

The use of IT during the 2004 hajj allowed the grand mufti of Saudi Arabia and chairman of the Council of Senior Religious Scholars to exert his global leadership role. He delivered a religious message against terrorism before an international body of pilgrims. On one level, the message was heard in Mecca and intended to be shared with other brethren when the pilgrims returned home. But because the message was televised, it was instantly diffused well beyond the borders of Saudi Arabia. This demonstrates how IT contributes to the expansion of Islam, allowing various international nodes to be in contact with the teaching of the highest religious authority of the kingdom. Moreover the tape-recorded sermon was also circulated through radio and videocassette for those who stayed at home. The direct circulation of sermons via television or the Internet and their indirect circulation through audio transmission and videocassettes exemplify the globalization of Islam as it reaches out to extraterritorial members of the faith.

Since 1999, the Internet has been an additional medium of global communication in the kingdom. The state server located at King Abdul Aziz City for Science and Technology processes incoming and outgoing messages, whereas the Ministry of Post, Telegraphs, and Telephones is in charge of developing and maintaining the infrastructure to censor the spread of forbidden material such as pornographic images or illegal transactions.[49]Each ministry of the government now has its own website and a large number of available pages, which can be accessed at www.saudiinf.com, with information on various aspects of Saudi Arabia.

A central website devoted to the hajj at www.tohajj.com, launched in 2001 by the governor of Makkah Province, provides information on the pilgrimage in several languages.[50] It is a source of information not only for pilgrims, but also for non-Muslims who want to enhance their knowledge of the faith. Although the Wahhabi version and tradition is prominent, it also addresses the traditions and interpretations of other Islamic schools of jurisprudence.

The advent of the Internet also has led to a new form of proselytism. Local imams in some European cities use the Internet for virtual interactions in which they respond to questions put to them online, to circulate information among the faithful, and to provide advice. These online imams are afforded a virtual channel through which they are able to interact with their audiences in cyberspace. Thus, a virtual *ummah*, like a digital diaspora, is becoming more and more an extension in cyberspace of communities of the faith.[51]

The Internet is used for official business as a mode of communication with overseas leaders, but it is also used informally among friends and colleagues to inform, seek advice, influence policy, and discuss matters of common interest.

Informing people about proper Wahhabi sites is another area in which the local community uses the Internet. For example, in Casablanca, one can find a bookshop not only selling books and beauty products from Saudi Arabia, but also displaying on its front façade the addresses of thirty-nine Wahhabi Internet sites that the owner recommends to clients and passersby.[52]

GLOBAL GOVERNANCE AND THE REMAKING OF NATIONAL GOVERNANCE

An unforeseen aspect of global governance is its unintended consequences for national or domestic governance. Through this process, national governance becomes a central node of global governance. The global governance of Islam is totally integrated in the practice of national governance. This is an important point, because one usually conceives of global governance as something above local governance, not as another face of local governance. In this perspective, the local and the global are seen not as poles on a continuum, but, as in the relationship between the nation-state and religion in Islam, as different faces of the same reality.

For Saudi Arabia to serve as an efficient host to pilgrims requires strategic planning for massive construction and maintenance of the necessary infrastructure. In the past forty years, the government has built several bridges, constructed multilane highways, and established water-supply reservoirs to ease the traffic of pilgrims and to make the country more hospitable for its guests. The two holy mosques in Mecca and Medina have also been enlarged to accommodate more pilgrims.

Because the obligation to make the pilgrimage is extended to both civilians and government officials, the Saudis must also ensure the security of these officials and officially welcome them to the kingdom. William Quandt remarks that "political leaders are obliged to devote considerable time to receiving heads of state, for whom the pilgrimage is good politics back home and a chance to seek Saudi aid with or without invitation."[53] These are opportunities for foreign visitors and pilgrims to discuss political matters of common interest with Saudi government officials.

Because the hajj is a major source of revenue for the kingdom, it must be managed to minimize cost and maximize profits. Before oil took over the leading role, the hajj was the main engine of the economy, due to the amount of money pilgrims used for ground transportation, lodging, and food. Even today, as Long remarks, the hajj is still "the premier commercial season in the kingdom, similar to Christmas in Western countries."[54]

The hajj also raises national security questions for the Saudis, who must be alert to prevent troublemakers from landing in the country, participating in the pilgrimage, and fomenting problems. Collaboration with foreign countries is necessary to identify and screen potential problematic pilgrims. In the recent past, Iranian activists have protested in Mecca in an effort to bring attention to the Saudi regime's shortcomings and failures. The capture of the central Mecca Mosque in 1985 by Iranian Shiite commandos was one example of how hajj security could be breached.

The global governance of the pilgrimage is predicated on the ability of the Saudi government to use it for a uniquely spiritual purpose, such as televising the rituals and the sermon of the grand mufti to the rest of the world at the end of the pilgrimage and distributing the Koran free of charge to pilgrims and for a strategic purpose, to achieve solidarity and unity among Muslims for a jihad to liberate Palestine, while at the same time preventing or dissuading dissidents from using it for revolutionary purposes, as happened on November 20, 1979, when affiliates of the Movement of the Muslim Revolutionaries of the Arabian Peninsula protested, or on July 31, 1985, when Iranian resisters demonstrated.[55]

Although there exists a hajj ministry in the government, other ministries are also involved in their own domain to accommodate pilgrims. The Ministry of Transportation oversees the mobility of ground transportation to Mecca and Medina, the Ministry of Health establishes clinics at times mostly staffed by foreign medical specialists hired for the occasion, the Ministry of Information is in charge of the satellite transmission of televised programs, and the Ministry of Foreign Affairs takes care of visiting foreign dignitaries, whereas the security services maintain their surveillance of suspected individuals.

Whereas hosting the hajj has consolidated the regime, it has also consolidated the position of some commercial Meccan families that hold a monopoly over the exchange and transactions of certain goods with the pilgrims.[56] These merchant families have been in the business for years and have established themselves as the main purveyors of key services by reason of past practices and political connections.

The analysis of distributed cellular network governance necessarily leads us to study the mode of connections between sites and to identify one or more sites that seem to occupy a more prominent position in the network than others. Mecca certainly seems to occupy a position resembling that of Jerusalem in Judaism or Rome in Catholicism. Seen in that light, the role of Mecca in the global governance of Islam, and of Wahhabi Islam in particular, deserves our attention. Just as Jerusalem is the geographical and spiritual center of Judaism, Mecca performs the same role for Islam. However, to unveil the hermeneutic contents of this holy site, one must analyze more closely various aspects of that centrality. Mecca symbolizes five global referents for Muslims living outside the kingdom. It is seen, first, as a point of origin that recognizes Saudi Arabia as the birth place of Islam; second, as a point of reference, because Muslims face Mecca five times a day during prayer; third, as a destination because of the obligation to make a pilgrimage to the site at least once in a lifetime; fourth, as a point of orthodoxy that the Saudis, through Wahhabism, use to ensure that the correct knowledge of Islam is produced, preserved, and protected from liberal Sunni, Shiite, and Western interpretive contamination; and fifth, as a point of global leadership, where the mufti preaches on the last day of the pilgrimage and communicates an interpretive message to Muslims around the world, via information technology, which is similar, in its global reach, to the "Urbi et Orbi" message and blessing of the pope.

Sites are not born Wahhabi, they are made Wahhabi. In the process of the religious transformation of a site, the school and the mosque are the two main arenas where indoctrination into Wahhabi Islam occurs in Saudi Arabia. A similar mode of religious socialization is also used by Wahhabi Saudis overseas. As a way of opening new Wahhabi enclaves abroad and injecting new blood into old enclaves, Saudis have established and financed mosques, sponsored schools, trained teachers in the kingdom, and assigned missionary preachers to foreign countries to consolidate the faith among the brethren and convert new recruits.

Network governance is not always a smooth process, because the national and global poles of attraction do not seamlessly converge in the same direction. In fact, there is an ongoing tension between the Islamic nation-state in its effort to influence the birth of a nationalist Islam—one that espouses a more state-centered ideology—and global Islam, which develops its *ummah* as a borderless nation or supernation that crosses the boundaries of nation-states. This tension between local and global expression of Islam is intrinsic to the network governance of Islam as such.

However, in Wahhabi Islam's efforts to function as a distributed cellular network, there is a tension between the governance of the internal network and its relations with other nodes that do not fall directly under its leadership because of different theological orientations. Unlike Judaism and Catholicism, in which a managerial clergy class choreographs linkages between congregations, in Wahhabi Islam, the concept of leadership is itself ambiguous because the religion has no room for intermediaries between the faithful and Allah. The relationship with Allah negates the existence of a priestly class. Therefore, extraterritorial linkages can be undertaken and spearheaded by any individual, group, or institution.

Global governance implies not only linkages with outposts outside the country, but also the development of a local infrastructure for the administration of the network. In the case of Saudi Arabia, this means, among other things, the creation of facilities (highways, housing, tent cities, medical clinics, ground transportation, and drinking water supplies) for hosting pilgrims to Mecca and Medina. In other Islamic countries, this function includes the establishment of an official government agency to manage the participation of individuals in the hajj. In other words, network governance requires the transformation of participating countries to meet the religious needs of global Islam.

Moreover, unlike situations in which church and state are separate and pursuing different agendas and the onus of developing infrastructures such as universities in which to train religious leaders falls on the church, in Saudi Arabia, where the state and Wahhabi Islam blend together and complement each other, the state is willing to pay for the infrastructure needed for the work of the official religion. The construction, expansion, and renovation of religious buildings exemplifies such undertakings by the state.[57] As part of its work, the government engages in long-term planning to provide security to the pilgrims, to ensure their well-being while in the kingdom, and to enhance the positive economic impact of this form of religious tourism in terms of the profit made by merchants thanks to the heavy flow of visitors.

CONCLUSION

Wahhabi Islam, in its deployment throughout the globe, uses a model of distributed cellular network governance that best embodies and reflects its various forms of expression in different localities and the active or dormant relationships among its nodes. Although Mecca serves as its birthplace and Saudi Arabia as a central bureaucratic site of leadership, most nodes maintain their autonomy despite their interdependence with the rest of the network. It is a loose global network that most intensely activates its trans-border linkages during the pilgrimage season and in moments of national crisis. However, some nodes maintain closer relations with each other than with the rest of the global network.

During the pilgrimage season, global networks of Wahhabi Islam are activated through the relations of Saudi Arabia with official agencies of foreign governments with a sizeable Muslim population in their midst. However, not all of these foreign agencies are proponents of Wahhabism. Furthermore, Saudi embassies and consulates deliver visas to prospective pilgrims and do not discriminate against those who do not follow this particular branch of the faith. Visas are provided on a quota system per country to those who apply and are sponsored by the local leaders of their mosque to ensure that non-Muslims and potential troublemakers are screened out in the process. Saudi Arabia maintains strategic control over the process of allowing pilgrims to come to the kingdom and managerial control over their activities while they are in the country. What, therefore, does distributed cellular network governance contribute to our understanding of this particular transnational religious and institutional context?

Wahhabi Islam presents a peculiar form of transnational informal networking of nodes: informal in the sense that linkages are established on a voluntary basis and, therefore, are not pre-scripted by religious law; transnational because Muslim diasporic nodes are for the most part multicultural and multiethnic in the sense that they comprise memberships who adhere to diverse Islamic traditions and immigrants from different countries. A mosque in a diasporic neighborhood attracts diverse members of the faith because it may be the only Islamic place of worship available in the area. That is why naming a node in North America or Europe a Wahhabi site can be misleading: such a qualifier may not, or does not, take into consideration the plurality of religious practices of offshoot multiethnic congregations.

One may then speak of different modes of relationships among nodes and between Saudi Arabia and other sites. These relationships may be activated depending on contextual circumstances, history, and needs; the very notion of distributed cellular network governance implies a diversity of nodes, a diversity of relations between nodes, a diversity of network arrangements and engagements, and a diversity of outcomes.

As a result of being the custodian of the holy sites of Mecca and Medina, and the dependence of the faithful on the kingdom for the annual pilgrimage, Saudi Arabia has positioned and projected itself as the interpreter par excellence and maximum leader of Wahhabi Islam. In matters related to the cellular network governance of the faith, here again Saudi Arabia occupies an enviable position. However, because no site totally controls the practice of the religion, various nodes have emerged to influence the direction of the faithful. In this sense, one may conclude that distributed cellular network governance implies a dispersed, mobile, and variable leadership structure because any node has the opportunity to intervene in the process, because key aspects of network governance depend on the willingness of different entities to contribute to the implementation of rules and partake in the processing of tasks, and because the composition and shape of power geometry may take different forms and characteristics at different times. All in all, the distributed cellular network governance model calls our attention to the malleability and variability of the multiple expressions of Wahhabi Islam throughout the globe. It further makes evident the mechanisms of intervention, collaboration, and accommodation the network has generated and put to good use for the success of its enterprising proselytism, humanitarian practices, and the flourishing of the faith all over the world.

5 The Network Management of Global Religions

Each of these monotheistic religions—Judaism, Catholic Christianity, and Islam—projects itself as a global faith with a cosmic dimension in terms of its spiritual content (the practitioners worship the God of the universe), universal membership (brethren live in every country throughout the globe), multinational expansion (in several countries, new recruits through proselytism have joined the movement), helping others wherever they might be located (the faithful provide aid to those in need in the Global South and the Global North), and with a message that is global (it is addressed to men and women of goodwill irrespective of their place of birth, race, class, age, gender, or sexual orientation).

However, the three religions define the global scope of their domain of governance differently. The theological doctrine (Torah, Bible, or Koran), the legal framework (halakha, canon law, or shariah law), and the composition of the global membership (kehillah, *ecclesia*, or *ummah*) each influences in its own way the trajectory and form that network governance and management may take in the context of a specific period, in a specific region, and for a specific group of people. This concluding chapter will analyze how these variables have shaped global governance for each faith.

GLOBAL RELIGION AS A MULTINATIONAL ORGANIZATION

As religious formations have become multinational organizations through globalization and the geographical expansion or creation of satellite communities,[1] they have developed multiple bureaucracies with their own administrative specificities, political environments, and cultural organizational practices. Governance in this context, with or without a headquarters, is concerned less with the micromanagement of each of the nodes and more with the coordination of the network and the use of its resources. This is so because the nodes are diverse, their resources are unequal, and power imbalance is a constant impediment to the equilibrium of the network.

Each multinational organization has its own life cycle or trajectory.[2] It is prone to extension as new nodes are added and contraction as nodes disappear because they are no longer viable institutions, which may be the result of secession, annexation, migration, or natural disasters. The network must manage the growth and decline that constitutes its totality.

Managing power imbalance between nodes is a central feature of network governance because it implies the restructuring of some units, the sharing of resources, and mobility within the network. In global religions where the units are not locked in step with the headquarters, each unit maintains its autonomy while contributing to the network. Despite that autonomy, there are often examples of codependency. One such example is that when selecting a rabbi or imam, it may be necessary to consult with the network to find the appropriate candidate. The issue is not whether or not a multinational religious organization functions as a hierarchy, but rather whether the network collaboration, dependence, or relationship is minimal or maximal.

A central feature of a multinational religious organization in the decision-making arena is that the composition of the number of actors varies.[3] Narrow or broad global participation depends on the nature of the problem to be addressed, which may affect the flexible size of virtual networks of participants. The diversity of practices may also vary due to the way in which some nodes use digital networks intensively, as in the case of those in North America, whereas others do so more selectively or very little because of a lack of or limited computer access.

The distinction between the Catholic Church and the other global religions is that the corporate headquarters of the former is called upon to manage sets of networks instead of governing distinct religious, pastoral, and humanitarian projects. In the case of Islam and Judaism, different sets of networks govern themselves and at times coordinate their actions with other networks. In the Catholic Church, this is done to uphold the theological orientation of the faith and to follow canonical procedures, but in the others, it is done for a double purpose: to maintain the autonomy of the nodes and to rescue or protect the larger body of the believers of the faith, including their national or ethnic groups.

LOYALTY TO THE NETWORK INSTEAD OF EXCLUSIVELY TO A SITE

Employees tend to be loyal to the firm that hires them. In the networked firm, loyalty is also to the network, and not exclusively to any specific site. Global religion as a network organization exhibits this fundamental feature of the networked firm. Loyalty to the network in the case of global Catholicism is seen in the way in which the mobility of seminarian students and clergy is handled. They may be sent to an overseas training ground so that they can be available to serve wherever they are needed. In fact, they may be appointed to serve based not simply on their preference, but also on where their talents can best be put to use. Loyalty to the network is also a genuine feature of global Judaism and Wahhabi Islam. For example, American rabbis have been hired to lead synagogal congregations in Europe. Likewise, Egyptian religious leaders have also been appointed to head the Grand Mosque in Manhattan. The advantage of network loyalty is that the most valuable actors can be identified to serve anywhere in the network. The network provides an opportunity to move from one site to another for travel, schooling, work, and cooperation.

In this environment, the religion is experienced as a global entity and mobility can be affected within the nation-state, as well as beyond its borders. In the Catholic Church, the pope has no territorial limits in the free democratic world in terms of where he can assign a priest or bishop. The other religions enjoy similar freedom in the global movement of their religious leaders.

Loyalty to the network makes it easier for a religion to move a religious leader in trouble to a different site in the network, which facilitates the resolution of local conflicts. Loyalty to the network can also reduce costs, such as when the religion uses the free services of its expert members. The Vatican consults such experts in Italy and abroad when they are asked to sit on papal commissions, and the local bishop consults professionals in his own diocese and beyond for construction, management, financial, and legal advice. These experts usually serve on a pro-bono basis or for reduced fees. Judaism and Islam enjoy a similar kind of access to experts and services among the professionals that comprise the networks of practitioners.

OUTSOURCING SPECIFIC FUNCTIONS

Unlike commercial firms that outsource work to any competent organization that can lower production costs,[4] the Catholic Church outsources programs to any appropriate node in the network. Outsourcing has been in operation in global religions long before it was used as a revenue-maximization and cost-minimization strategy in the modern firm. Outsourcing is the operation by which a firm allocates work to a satellite or outsider firm (or even to a freelancer) for the purpose of completing it on its behalf.

In addition to direct outsourcing, there is also an indirect form of the practice. It occurs when a node establishes a program for the benefit of the network and requests the headquarters to support the effort. Such sponsorship transforms the effort into an official operation outsourced to a particular node. An example of such an indirect outsourcing form is the Catholic Middle East Program. Although it was created in the United States, in order to be considered an initiative of the global church and not simply a local arrangement by a local node, it had to be sponsored by the Vatican. Such a strategy was needed to help resolve issues of logistics, including fundraising and program implementation. Because the Vatican was not keen about managing the initiative itself, it outsourced the program to the diocese of New York. The rationale is that the archdiocese is asked to share its wealth with some nodes of the global network and to take the lead in terms of influencing American and Canadian foreign policy toward the embattled Christians and Catholic patrimony in the Middle East. In this case, the Vatican oversees but does not micromanage the process.

In Judaism, outsourcing takes place in countries that have a chief rabbinate, such as England and France, and within the Hassidic movement. A central authority or the rebbe sometimes asks a synagogue to help another in the network that is under duress. This process is similar to the experience of the Catholic Church. In addition, some Jewish congregations throughout Europe outsource the task of securing synagogues on shabbath to security personnel management services. In Saudi Arabia, because Islam and the government are part of a common project, during the hajj some functions such as medical care for pilgrims are outsourced to foreign medical personnel. In the context of the administration of global religion, outsourcing takes different forms related to the existence or absence of a corporate headquarters.

NETWORK AS INNOVATION INCUBATOR

The network provides an environment for the incubation of innovation. Because of the different positions of nodes, some are more prone to experiment with new things than others, whereas some are more prone to deviate from standard practices. Although the network upholds the mainstream line, it may incorporate initiatives carried out by the nodes. An experiment that started in Spain, and then in France, for example, which invites parishioners to pay a stipend to the local priest to pray for their intentions at Mass, was first rejected by the church headquarters and then incorporated by the Holy See as a mainstream practice of the Church.

Innovation comes about because no node controls the production of doctrine, interpretation, and traditions. Good practices in one place may be appropriated for the benefit of the global network. This is seen in the canonization of saints, who were virtuous persons elevated to the status of sainthood. As such, they belong no longer exclusively to the local parish, but also to the global church.

The network provides niches for innovation because it allows exchanges of ideas between individuals and groups, facilitates searches for solutions to common problems, tolerates both competition and collaboration, and provides incentives for people to develop new ideas, new programs, new interpretations, new analytical frameworks, and new practices. After Vatican II, local churches were very much involved in innovating new mechanisms to renew the global network. The Dutch church experimented with the celebration of the Mass at home, the Latin American church with liberation theology, the American church with sacred choreography, the Caribbean church with traditional musical instruments, and the African church with the incorporation of folk religious rituals into the Catholic liturgy. These experiments were undertaken for the purpose of renovating old structures and recalibrating old practices of the global Church.

Judaism, Catholicism, and Islam have each used the network as an innovation and diffusion incubator in matters pertaining to religious architecture, as a model of a synagogue, church, or mosque developed in one site may be implemented or copied elsewhere; scholarship, as discussions among pairs may lead to new interpretations; and liturgical practices, as new forms of worship used in one site may be experimented with to meet the contemporary needs of the faithful.

MANAGING RELATIONS

Network governance requires the management of three sets of relationships that may enhance or impede the functioning of the system. The relations that each node maintains with the network are crucial for the operation, stability, and reproduction of the system. This constitutes the central feature of network governance and management. The relations between nodes are a second important feature. If there is any conflict between them, this may disturb the entire network. They must be managed by the network to prevent any disruptions of its activities or general collapse. In other words, local crises are matters of importance to the network and must be resolved to prevent them from expanding and paralyzing the system. The relations between the network and external networks constitute the third important feature in network management. These relationships are maintained for the purpose of collaboration so as to entertain a modus vivendi between competitor faiths and also to protect the borders of the religion to prevent confusion and external encroachment on their specific spheres of action. The three religions all attend to these relations differently and have established different protocols to deal with issues pertaining to both internal and external networks. The legal systems they have developed provide the means by which they resolve internal and external conflicts.

However, the management of human resources and personnel constitutes a mechanism that is common to all three faiths. This involves the boundaries of tasks, the yearly manipulation of wages to meet workers' expectations, the recruitment, dismissal, or retirement of personnel, and the establishment of relations of authority as set by religious law. These bundles of relations play out differently in different locales, so the global management of administrative diversity is also an issue that requires attention. In Judaism, a parallel structure of lay trusteeship exists to process these activities, however in global Catholicism, the role of the laity is subservient to the priest and bishop and must be seen foremost as a dependent structure. In Islam, managing relations is carried out in a nongovernmental sponsored mosque by the imam in charge of leading the Friday prayer, and in government sponsored mosques, some form of trusteeship is put in place. The role of the trustees is to provide guidance to the imam and his collaborators and in some cases to intervene if a leadership replacement is needed.

NETWORK MANAGEMENT WITH OR WITHOUT HIERARCHY

As we noted, network governance does not necessarily eliminate hierarchical governance. Hierarchy can and does exist side by side with network governance. The three religions proceed with these twin forms of governance. Although hierarchy is more pronounced in the case of the Catholic Church, it is practiced in Judaism at the synagogue level with the figure of the rabbi, and in relations to the headquarters of the movement in the case of the Hassidim. Although Islam does not recognize the need for intermediaries between human beings and Allah, the role of the imam in the West and the ulama in the Middle East reflects some forms of operational hierarchy in the management of the mosque and the faith.

Hierarchy or ranking is seen not only in the operation of the nodes, but also in the deployment of the global network. At the node level, ranking is practiced for efficiency and procedural purposes. In the Muslim case, this is more nominal than substantive. In contrast, a global network is likely to include nodes with unequal power, wealth, and influence, as well as nodes with mature or institutionally incomplete constituents. Although hierarchical ranking may not be formally institutionalized, in everyday practice it is obvious that some nodes are better positioned in the network than others.

In the Catholic Church, the upper echelons of the hierarchy constitute a power elite network and occupy a position of global leadership that is unmatched at the level of lower echelons. In contrast, neither Islam nor Judaism displays such a controlled network hierarchy structure because both lack a centralized bureaucratic management.

INFORMATION CREATION, CONSUMPTION, AND DIFFUSION

The network creates information that it then consumes and diffuses. Because the network has its own identity, it creates the information that would sustain that identity. Individual scholars or leaders, specialized bodies, and group meetings produce knowledge for the network to consume. This transglobal mechanism allows the faithful to evolve at the same time or to be on the same page as the leaders. This produces public knowledge that can be discussed, challenged, and refined. To protect its doctrine, the group decides what is orthodox and what is not.

The role of nodes in the production of religious knowledge varies. Some produce more for local consumption, whereas others tend to produce more for consumption by the network. This feature distinguishes the local parish from the Vatican in information diffusion, for example.

Information creation, consumption, and diffusion are also managed by the network and are difficult tasks to accomplish because of the number of people involved in these processes, the need to do so to protect the integrity of the network, and the complexity involved in arriving at some consensus in order to sustain the faith of the membership.

Information diffusion must be managed by the network to control content and to absorb the costs involved. The three faiths use various mass media to do so, including books, pamphlets, audiovisual devices, radio, television, and the Internet. The global management of nodes that spearhead such an effort constitutes another layer of the global governance of these faiths.

MANAGING TURBULENCES IN THE NETWORK

A global network may be perturbed by crises in any of the nodes.[5] During these periods, it becomes imperative to govern the local crisis so as to help the site recover from its problem and also to prevent it from disrupting other nodes. Because the religion is worldwide, any node may find itself in this predicament, not because of what happens internally in its organization, but rather because of external factors such as political instability, natural disasters, and other variables over which it has no control. The network serves as a shock absorber for these nodes, nurturing and nursing them back to a healthy state.

A central node may get involved in the resolution of such a crisis, however it is not always necessary. Another node or a group of nodes may come to the rescue of the one that is experiencing difficulties. Thus, the elasticity of the network allows for different forms of intervention by different groups of actors.

Although turbulence from external factors cannot be completely prevented, the network provides flexible ways for nodes to intervene and resolve issues. Turbulence may also consolidate ties between nodes by reprogramming them, reshaping the network, and resolving conflicts experienced by the network or any of its constitutive nodes.

THE NETWORK MANAGEMENT OF LOCAL UNITS

There is no uniform way to globally manage a local node, because doing so depends on how the node became linked to the network and the history of the node.[6] There are three ways in which local units are attached to the network. They may be the outcome of a process initiated by the headquarters, as in the case of the establishment of mission churches. These are created for the purpose of proselytizing, and it is understood that the religious authorities will be of a foreign stock. The second form of attachment is when a national government takes the initiative to invite the missionaries to establish nodes. This form of contractual entry may imply gifts of land for the establishment of the place of worship, financial contributions by the state, a role for the government in the appointment of religious leaders to certain posts, or the expectation that the missionaries will run schools and hospitals in addition to their religious apostolate. The third mode of attachment is through annexation. This occurs when a congregation joins another church. For example, in the recent past, Anglican congregations have joined Catholic dioceses because of dissent over doctrinal and moral issues.

These three models of establishment of a foreign node require different managerial interventions. For example, whereas the needed resources (money and clergy) must be collected or assembled for the first model to succeed, the last requires less of an intervention, because there is no need to create a national structure. In the second model, the collaboration of the national government makes the effort less financially demanding.

As an integrated system, the links that characterize the working of global religions are not simply between the management of the headquarters and the management of satellite sites, but also between the rank and file: people to people, associations to associations, and congregation to congregation. For example, as has been noted repeatedly, local churches, synagogues, or mosques in the Global North have come to the assistance and rescue of local Catholic, Jewish, or Muslim congregations in the Global South.

There comes a time when expatriate missionaries no longer control the hierarchy of the local religious leaders. At first, expatriates are needed for capital, expertise, and the ability to communicate with the headquarters, but since Vatican II, the trend in the Catholic Church has been to replace expatriates with local priests. This is because of the structure of the network: leadership can be exerted by any individual node or by the best qualified in the network because the network belongs to all of its units. Although in the Catholic Church the corporate hierarchy oversees the management of local units and the extent to which they comply with canon law regulations, in Judaism the local congregation decides on the management of its affairs and on whether or not it wants to keep or fire a rabbi. The network reappears in the hiring of a rabbi because he or she will be recruited from the network. In Islam, the congregation manages its own affairs, in some cases with the oversight of trustees. A node in the network may intervene to help financially and with logistics.

The integration of the network comes about because lessons learned in one node can be duplicated elsewhere, talent from one node can be transferred elsewhere, associations established in one node can be implemented elsewhere, liturgical practices developed in one node may be used by the network, and financial resources from one node can be shared with others. These various levels of interaction, appropriation, borrowing, and collaboration reflect in no ambiguous way the integration of the network.

THE CHANGING STATUS OF NODES

Nodes do not necessarily have the same capacities or maintain the same positions during their lifespan.[7] They may become stronger or weaker in the network because of migration, the ascendancy of leaders, or wealth. For this reason, network management has a variable shape depending on the ebb and flow of nodes. Network governance and administration are mobile activities because of the changing positions of nodes in the network.

This constant making, unmaking, and remaking of the network brings life, new perspectives, and new resources to the global circuit. It does so by repositioning the nodes. For example, Berlin's main synagogue has lost the preeminence it once had in global Judaism prior to World War II because of the Holocaust, which decimated its membership. Istanbul has also lost its preeminence in global Islam because of the collapse of the Ottoman Empire. The same thing happened to Rome during the period when the papal bureaucracy was relocated to Avignon.

The status of a node may also change because of the migration of a leader to another site. One may think here of the role of Brooklyn as the site of the headquarters of the Lubavitch Hasidic Movement due to the emigration of the leadership from Eastern Europe to New York. One may also refer to Israel, which has regained its position of global leadership in Judaism since its establishment in 1948. In this light, one may venture to say that the architecture of the global network is likely to change as a result of the changing status of nodes.

THE GLOBALIZATION OF THE MANAGEMENT OF THE FAITH

Network management evolves to meet different needs at different times, depending on the phase of development of the network. The change or shift is from centralized administration to multiple and decentralized management.

At a deeper level, network governance requires a model of territorial organization that reflects the spatial effect of religion on society. Network governance implies not simply governing people, but also managing a form of territorial practice. In fact, the efficacy of network governance cannot be understood outside this territorial logic. This is why it must be said that global governance implies the transnationalization of this territorial logic. Depending on the strength of the religion, this spatial logic may be imposed or constructed in the process of territorial integration.

One may then speak of "territorial recomposition" or "territorial construction," which provides the global religion with a spatial infrastructure for the operation of its institutions.[8]

The ultimate goal of this spatial logic is to provide territorial infrastructure for both national and global governance so that one will not handicap the other.[9] Network governance works at the national level to ensure cooperation among brethren and at the global level to reflect the heterogeneity and mutual dependence of the nodes. This transnational, religion-based territorial structure and logic evolves over time and depends on whether the religion expands or contracts.

MULTIPLE ADMINISTRATIVE CULTURES

Network management presides over diverse local administrations that have their distinct cultures and traditions.[10] The differences appear not in their bureaucratic forms, but rather in their forms of practice that are culturally based. Network management of the bureaucracy with its multiple forms and divergent practices is a daily challenge confronted by these global religions. The different practices of those who join the faith as already organized units with their own traditions may lead to such diversity. The mode of entering the network may be an important indicator as to whether they join the circuit as individuals or organized groups. As individuals, they may be more likely to follow the network, but as groups they may keep their traditions in the network.

Global religions have used different methods to establish local administrations in foreign countries, to interact with those that emerge from local initiatives, and to negotiate with the idiosyncratic administrative cultures of the Global South.

At first, the Catholic Church attempted to impose a single administrative culture on local bureaucracies everywhere—one bureaucracy with branches outside of Rome—with procedures spelled out in the code of the canon law. The advantage was that similar procedures for similar problems were followed everywhere. A judgment at a lower level could be thrown out on the basis of procedural matters, or a case could be sent on appeal to Rome for a final resolution. In a sense, the Curia functions as a supreme court for these hard-to-solve cases. Until Vatican II, this strategy totally ignored local administrative cultures and imposed a bureaucratic form influenced by the techniques of colonization, domination, and subjugation adopted by secular empires.

After Vatican II, the leadership of the Catholic Church began incrementally to select indigenous bishops instead of imposing foreigners on mission churches and to allow local bureaucrats to use their national language in the preparation and exoneration of canonical cases and in communication with the Curia. The Vatican has pluralized its corporate structure by appointing bishops from the Global South to Curia positions in the hope that they might be more attentive to local cultural concerns. This multicultural leadership is supposed to lead the effort in the management of the dispersed multicultural bureaucratic administration.

Despite these changes, the bureaucracy remains largely influenced by the centralized, hierarchical model that preceded it.[11] As long as canon law requires the use of same procedures and the existence of a similar bureaucracy throughout the world, local indigenous forms of bureaucracy will continue to be trumped by the hegemonic bureaucratic model, and the managerial role of the bishop will be to negotiate the outcome between pressure from Rome fuelled by the requirements of canon law, on the one hand, and the cultural needs of the local community (the diocese or archdiocese), on the other. This Latin model has been imposed even on the primitive Christian Church of Jerusalem, the portion that is in communion with the Roman Catholic Church.

The specificity of local administrative practices is more marked in global Judaism and Islam, where the local units have more autonomy because they are controlled by the membership, and not by an external entity. The result is a diversity of practices. For example, whereas managing the security of synagogues is a major concern in Central and Eastern Europe, this is a less acute issue in North America. One may also think of other differences in administrative practices: the divergence in synagogues that welcome female rabbis and those that do not, discrepancies between Sephardic and Ashkenazi styles of synagogue management, and the disparity between richer synagogues that have well-financed programs and those that function with the bare minimum. Islam provides a similar array of local administrative practices fueled by national cultural differences and whether or not they are sponsored by a government. For example, the administration of a village mosque in Egypt pales in reference to the elaborate administrative process of the Grand Mosque in Manhattan with the oversight provided by representatives of Muslim countries at the United Nations.

MANAGING NETWORK DIVERSITY

Governing global religions thus does not require any particular style of administration, but rather the ability to engage in the management of diversity in the network.[12] In the business world, this is seen when a firm borrows features from the bureaucratic model whereby an office may be in charge of the administration of a specific sector. Diverse sectors with different cycles of performance intertwine and must be managed at different speeds, because they display different rationales and require different approaches. The same applies to the management of diversity in the networks of global religions. One may think of the governance of the financial basis of the operation, with its collections of funds from the faithful, the investment of portfolios, the disbursement of expenses and overheads, and the redistribution of the money for charitable and humanitarian purposes. This requires attention to detail to prevent the collapse of the institution. This, in turn, is different from handling of human resources, which involves the mobility of religious leaders, the disciplining of theologians whose teaching and writings are at odds with mainstream orthodox practices, and the promotion and best use of talent and skills. This, again, is different from the management of church patrimony, which may become a routine, or from that of doctrine and liturgical practices.

Here again there are differences of practices when one compares the Catholic Church with Judaism and Islam. In the Catholic Church, the Vatican plays a central role in the management of network diversity because of its oversight and regulatory role vis-à-vis the network. In contrast, the local synagogue and mosque plays an active role at the local level to ensure adequate management of diversity by the network. The top-down approach of the Catholic Church must be contrasted with the bottom-up approaches of Judaism and Islam.

LEADERSHIP VARIABILITY

Network governance unveils the variability of leadership not only because of the participation of all the nodes in the process and the turbulences generated within or outside the network, but also because of the managerial strategies that a node may pursue.[13] In the history of the Catholic Church, for example, popes have emphasized different aspects of global governance because of temperament, historical context, or the perceived needs of the Church. Paul VI was concerned with the detailed administration of the Holy See and John Paul II highlighted the pastoral mission of the papacy, whereas Pius XII devoted a good portion of his time to doctrinal matters and Benedict XVI to the maintenance of traditions and restoration of orthodoxy.

In Judaism and Islam, leadership variability has occurred because of the prominence of nodes or the rise of a leader that commands the attention of the network. This was certainly the case with the rebbes of Hassidic communities in Eastern Europe and now with the Orthodox establishment in Israel. In Islam, the shift from the Caliphate to Saudi Arabia in regard to Wahhabi Islam is another example of leadership variability.

Leadership variability may emerge as a result of the importance of nodes in the network and the movement of key players from one site to another. The status of a node may be elevated when someone in the site is called upon to occupy an important position in the network. During the reign of Pope John Paul II, the Polish clergy and faithful had easier access to the Vatican than the Nigerian or Chilean clergy, for example. With the papacy of Benedict XVI, the German Catholic Church has somehow regained the visibility it lost as a result of the atrocities of the Nazi Regime.

The migration of the headquarters of the Lubavitch Movement from Eastern Europe to New York City has upgraded the role of New York vis-à-vis the rest of the network. Likewise, the return of Ayatollah Khomeini from exile in Paris elevated Iran's status vis-à-vis the rest of the global Shiite network, including followers in Syria, Palestine, Lebanon, and the diaspora.

Leadership variability also accounts for variability in the architecture of the network in relations to the positions of nodes, interaction between nodes, the status of the nodes, the flows of communication, the governance of nodes, geographical segmentation, and the formation and identity of cohorts. Here, variability comes about as a result of the mobile leadership structure of the network, which produces different forms of modulation, cadences, and interventions. From the standpoint of the network, leadership variability is the rule, not the exception.

ADMINISTRATIVE TRAINING

Administrative training to provide the leadership of a node or the network varies from one global religion to another.[14] In all of them, administrative training is not considered a priority, when compared with religious training, and as a consequence, traditional methods prevail in the management of these units. One observes a lack of administrative preparation for priests and bishops, rabbis, and imams.[15] After graduate school in theology and some pastoral training in homiletics, priests or rabbis may be hired or appointed to serve a congregation. To administer the unit, they may get some help from a parish council or board of trustees, but that is not enough to manage large sums of money, large congregations or groups of parishes, and complex human resources. There is a need for the creation of specialized programs or schools that focus on parish, diocese, synagogue, and mosque administration, financial and human-resources management, and religious entrepreneurship.

This is a delicate issue. On the one hand, one does not want to transform religious leaders into administrators first, and priests, bishops, imams, or rabbis second. On the other hand, some professional training is necessary to ensure adequate financial responsibility because of the money, human resources, and infrastructure these leaders are called on to manage. When I asked a Catholic bishop how much administrative experience he had before coming to his diocese, his answer was, "none, formally," except his practical experience of serving as the pastor of a local church in another state. He said that overnight he was given a set of keys, a diocese to lead, and a large sum of money to administer.

NETWORK LEADERSHIP ROTATION

A central mechanism of network governance is the rotation of network leadership. The three faiths have developed three different forms of global leadership to energize their memberships. This leadership rotation is undertaken not only for matters related to doctrine, but also to convey a positive image of the faith to outsiders, and occurs as a result of the dynamic generated within the global structure of operation. In the case of the Catholic Church, for example, candidates from different dioceses and countries have been selected Pope during the conclave period—from Italy, in the case of Paul VI, Poland, in the case of John Paul II, and Germany, in the case of Benedict XVI. Each has had a different influence on the composition and the deployment of the global leadership network through the bishops and cardinals they promote and those they appoint to Curia positions or to serve as top religious leaders in their respective countries.

In Judaism, one node may step forward to provide global leadership. For example, five weeks before the visit of Pope Benedict XVI to the United Nations in April 2008, Arthur Schneier, the senior rabbi of Park East Synagogue, located on East 67th Street in Manhattan, was at the Vatican to urge the Secretary of State to add one more stop to the papal itinerary and to visit this Jewish place of worship and meet with a group of rabbis.[16] Here is a case of a node taking a leadership position on behalf of all the branches of Judaism in matters concerning the relations of the faith with the Catholic Church.

Islam presents yet a different form of leadership rotation. Its model of leadership rotation is more related to the political strength of a regime or group of international actors. The role played by Turkey through the caliph within the context of the Ottoman Empire was an example of a country-specific leadership, as is the role played by Saudi Arabia in the case of global Wahhabi Islam today.

DIGITAL MANAGEMENT

The entry of the network management of global religions into the digital age has been incremental and punctuated by a series of phases, from the introduction of the telephone as a means of communication only in times of crisis, to the facsimile to submit or exchange documents, to the television (which evolved into satellite television and now digital television) to deliver religious programs or relay religious events taking place thousands of miles away to local audiences, and, more recently, the Internet for inter-site interactions, to make religious information and documents available on websites, and facilitate network coordination.

Digital network management reflects the doctrinal orientations and traditions of a global religion. Because of these differences, there are unique ways that each global religion makes uses of digital technology. For example, in the Catholic Church, because of its bureaucratic arrangement and the canon law that establishes procedures of practice, the volume of digital communication tends to go usually from the leadership of a diocese to that of another diocese abroad, or from the Vatican to an archdiocese or to the apostolic nuncios and vice versa; whereas in the cases of Islam and Judaism, the initiative for global communication may come from any member of the faith or the congregation.

All the faiths use digital technology as a tool for global and network management, a means to expand the membership of the religion, a way to nurture practitioners, an instrument for global coordination, a method to enhance bureaucratic production and efficiency, a mechanism for fundraising and community exchanges, and a procedure to integrate the religion into the nooks and crannies of modern culture. With the advent of digital technology, each religion has positioned itself to use technology not only to advance its own agenda, but also to shape the moral dimensions of its deployment and its projected trajectory.

CONCLUSION

The globalization of Judaism, Catholicism and Islam creates the conditions for the emergence and sustenance of the network model of governance of these global religions. The network form of global management becomes an intrinsic feature of their administrative operation because it depicts not simply governance of a central node or at a distance, but also the governance of a transnational network of local sites. It is furthermore necessary because of the mobility, variability, and interdependence that characterize the nodes, which feed the malleability and flexibility of the global network. Such a transboundary network serves as a channel through which the faith choreographs its local identity and public form of expression.

Although globalization indicates the expansive nature of the network, the faith and sites involved distinguish the content, direction, and intensity of network flows. The imposition of a spatial logic—to the extent the religion establishes sites, feeds their interdependence and connections, and sustains their growth and expansion—that re-territorializes nation-states and the global landscape by the transnational network of its nodes, is another characteristic feature that distinguishes a global religion from another in matters of network governance. The logic of this transnational space is itself based on the behaviors of dispersed nodes that constitute the transborder infrastructure of network governance.

Network governance implies a variety of leadership roles, managerial tasks, and entrepreneurial initiatives and projects spearheaded by the diverse sites that compose the network. The ebb and flow of this transnational architecture of network governance is shaped by the multiplicity of network managerial interventions to expand, contain, or resolve network problems. In other words, the ability of different nodes to contribute to the network governance process shows the capacity of the network to absorb these interventions and explains why network leadership is a collective undertaking.

Throughout this analysis, three models of network governance espoused by Judaism, Catholicism, and Islam are spelled out to explain their characteristics and to distinguish their mode of operation from one another. The *cosmonational network governance* used in Judaism highlights the role of ethnicity and religion in the social parameters of the leadership and administration of the religion; the *corporate network governance* employed by Catholicism indicates the prominent role of the Vatican in maintaining the capability of diverse nodes to intervene in the management of the global church; and the *distributed cellular network governance* that is the hallmark of Wahhabi Islam recognizes both the autonomy and interdependence of nodes as well as the privileged role played by the Kingdom of Saudi Arabia as the custodian of the religious sites of Mecca and Medina.

However, unlike the network governance of the modern global firm that is heavily influenced by the principles of revenue maximization, cost minimization, and profit maximization for reasons of competition and survival, the management of these faiths is imbued with the religious values and beliefs they propagate. Their management must be gauged not by their efficiency to make a profit, but rather by their ability to proselytize, develop, communicate, and sustain the practice of the faith on a global scale.

All in all, it becomes evident in following the ramifications of this research that an analysis of the infrastructure and mode of network governance of global religions is an important investigation to undertake in order to understand the behaviors and operation of each faith throughout the globe. Network governance allows us to identify various forms of collaboration and tension among nodes, manifold interventions to maintain the stability of the network, the translation of the belief system in institutional life, and the role of the network in nurturing the lives of the dispersed brethren. Network governance of global religions—whatever form it may take—is a transnational institutional expression and channeling of the faith in the daily life of the global community of believers and practitioners.

The analysis unveils various aspects of network governance theory and provides a novel and alternative way of understanding authority structures and systems of religious communities in the global and digital age. Judaism, Catholicism, and Islam have been singled out for investigation and this approach can be applied to the study of other religions as well. The network governance approach explains the contribution of the network to the sustenance and viability of each node, the contribution of each node to the successful operation of the network, how some traditions of the nodes filter through to become part of the mainstream traditions of the religion, and how the understanding of the theological tenets of the faith by the nodes influences the content and shape of the doctrinal teaching of the faith and its dominant interpretation. Network governance theory sheds light on and provides a justification for the rationale of the evolution of doctrine. It indicates why an understanding of a religious doctrine based on the contribution of the network is bound to evolve over time.

Network governance theory places an emphasis on the identification of nodes, the relations between nodes, the relations between subnetworks of nodes, the weight and structural positions of nodes in the network, the intensity of relations between nodes, the impact of new nodes as well as the extinction of old nodes on the network, the node as a bundle of networks, the cross-border expansion of the network, the impact of nodes on the network and vice versa, and the formation of networks. Network governance theory enables the deconstruction of authority structures within the religious systems and their reconstruction in terms of networks. This insinuates unveiling the role of nodes in the meaningful governance of a religion. The approach identifies the social infrastructure of the organization and the mechanisms that sustain the cross-border decision-making system of the religious community.

Unlike the top-down approach of previous studies of authority structures of religious communities, which assume the existence of a hierarchical structure, the network approach focuses on the architecture of social relations between nodes to understand how a religion governs itself. It also pays attention to the belief system that shapes the contents and contours of these relations. In other words, what is expounded through this analysis is not that the globalization of these religions induces the faithful to develop a network governance apparatus; rather, it is the theological content of the faith, and historical developments, including the diasporic condition of members, that establish the framework for the adoption of a network model of governance.

Reading religious traditions through the lens of network governance conveys new ways of understanding global religions, including the realization that each global religion has been shaped by multiple actors in diverse countries. Think of the contribution of Muslim theologians in Turkey, Syria, Egypt, Iran, and Saudi Arabia to the interpretation of Islam and its moral principles; the role of Jewish religious leaders in Babylonia, Israel, and Europe in the shaping of Judaism as it is practiced today; and Catholic theologians including the Church Fathers from the Middle East, North Africa, and the Mediterranean Region so crucial in establishing the books included in the New Testament and the corporate network governance of the Catholic Church. The network governance model permeates many aspects of the organizations of these religious communities. It is unlikely that the future of global religions will differ from the network mechanism identified above, one which has existed since their inception.

Notes

NOTES TO INTRODUCTION

1. Mircea Eliade, ed, *The Encyclopedia of Religion* (New York: Macmillan, 1987), vol. 11, pp. 327–354; F. E. Peters, *Jerusalem and Mecca: The Typology of the Holy City in the Near East* (New York: New York University Press, 1986).
2. H. Busse, "The Sanctity of Jerusalem in Islam," *Judaism* 17 (1968): 441–468; S. D. Goitein, "The Sanctity of Jerusalem and Palestine in Early Islam," *Studies in Islamic History and Institutions* (Leiden: Brill, 1966), 135–148.
3. R. J. Zwi Werblowsky, "Jerusalem: Holy City of Three Religions," *Jaarbericht ex Oriente Lux* 23 (1973–74): 423–439; Emmanuel Sivan, "Le Caractére Sacré de Jerusalem dans l'Islam aux XIIé-XIIIé Siécles," *Studia Islamica* 27 (1967): 149–182.
4. E. E. Urbach, "The Heavenly and the Earthly Jerusalem in Rabbinic Thought," in *Jerusalem Through the Ages* (Jerusalem: Israel Exploration Society, 1968), 156–171; Sh. Talmon, "The Biblical Concept of Jerusalem," in J. M. Oesterreicher and A. Sinai, eds., *Jerusalem* (New York: John Day, 1974): 189–203.
5. Silvio Ferrari, "The Holy See And the Postwar Palestine Issue: The Internationalization of Jerusalem and the Protection of the Holy Places," *International Affairs* 60, no. (1984): 261–283.
6. Khaldoun Samman, *Cities of God and Nationalism* (Boulder, CO: Paradigm, 2007).
7. For an erudite and in-depth analysis of the evolution of Jerusalem and Mecca, see Peters, *Jerusalem and Mecca*; for the evolution of each site—Jerusalem, Rome and Mecca—as a nationalist project, see Samman, *Cities of God and Nationalism.*
8. J. S. Birks, *Across the Savannas to Mecca: The Overland Pilgrimage Route from West Africa* (London: C. Hurst, 1978), 1; Mukhtar Bhatti, *Lahore to Mecca by Road* (Lahore: Bhatti Publications, 1981), 18.
9. Michael N. Pearson, *Pilgrimage to Mecca: The Indian Experience 1500–1800.* (Princeton: Markus Wiener, 1996), 11.
10. Ghazy Abdul Wahed Makky, *Mecca: The Pilgrimage City* (London: Croom Helm, 1978), 23.
11. Matthew 16: 18–19: " You are Peter, and on this rock I will build my church . . . and I will give you the keys of the kingdom of heaven, and whatever you bind on earth will be bound in heaven, and whatever you loose on earth will be loosed in heaven."
12. Yves Congar, *L'Eglise* (Paris: Editions du Cerf, 1970).
13. Michel S. Laguerre, *The Informal City* (Basingstoke: Macmillan Press, 1994).

NOTES TO CHAPTER 1

1. On the historical evolution of these holy cities, see Khaldoun Samman, *Cities of God and Nationalism* (Boulder, CO: Paradigm, 2007); Brannon Wheeler, *Mecca and Eden: Ritual, Relics, and Territory in Islam* (Chicago: University of Chicago Press, 2006); Hamza Bogary, *The Sheltered Quarter: A Tale of a Boyhood in Mecca* (Austin: Center for Middle Eastern Studies, University of Texas at Austin, 1991); J. S. Birks, *Across the Savannas to Mecca: The Overland Pilgrimage Route from West Africa* (London: C. Hurst, 1978); and David Commins, *The Wahhabi Mission and Saudi Arabia* (London: I. B. Tauris, 2006).
2. Karen Armstrong, *Jerusalem: One City, Three Faiths* (New York: Knopf, 1996).
3. Missionary work is less common in Judaism; it is done mostly among Jews who have abandoned the faith in an effort to bring them back to the fold. In the case of Islam, see As'ad AbuKhalil, *The Battle for Saudi Arabia: Royalty, Fundamentalism, and Global Power* (New York: Seven Stories, 2004).
4. Thomas J. Reese, *Inside the Vatican: The Politics and Organization of the Catholic Church* (Cambridge: Harvard University Press, 1996).
5. William Garrett, "Canon Law and Corporate Church Action," *Virginia Review of Sociology* 1 (1992): 137–148.
6. Silvio Ferrari, "The Holy See and the Postwar Palestine Issue: The Internationalization of Jerusalem and the Protection of the Holy Places," *International Affairs* 60, no. 2 (1984): 261–283; Mark R. Kowalewski, "Firmness and Accommodation: Impression Management in Institutional Roman Catholicism," *Sociology of Religion* 54, no. 2 (Summer 1993): 207; Ralph Capio, "The Papacy: A Case Study in Organizational Longevity," *Journal of European Studies* 26, no. 4 (1996): 437–451; Y. M. Hilaire, "Note sur le recrutement des éleves du Séminaire Français de Rome (1853–1914)," *Archives de Sociologie des Religions* 12, no. 23 (1967): 135–140; Dean R. Hoge and Aniedi Okure, *International Priests in America* (Collegeville, MN: Liturgical Press, 2006); Thomas J. Reese, *Archbishop: Inside the Power Structure of the American Catholic Church* (San Francisco: Harper and Row, 1989); Walf Knut, "The Church Tax as a Means of Subsistence," in William Bassett and Peter Huizing, eds., *The Finances of the Church* (New York: Seabury Press, 1979), 20–27; Francis Oakley, "Constitutionalism in the Church," in Francis Oakley and Bruce Russett, eds., *Governance, Accountability, and the Future of the Catholic Church* (New York: Continuum, 2004), 76–87; and Thomas Reese, *A Flock of Shepherds: The National Conference of Catholic Bishops* (Kansas City, MO: Sheed and Ward, 1992). See also Nasser Ibrahim Rashid and Esber Ibrahim Shaheen, *Saudi Arabia: All You Need to Know* (Joplin, MO: International Institute of Technology, 1995) and Abdulla Mohamed Al-Zaid, *Education in Saudi Arabia: A Model With Difference* (Jeddah: Tihama, 1982).
7. Mark Newman and Albert-Laszlo Barabasi, eds., *The Structure and Dynamics of Networks* (Princeton: Princeton University Press, 2006); Albert-Laszlo Barabasi, *Linked: The New Science of Networks* (Cambridge: Perseus, 2002); S. N. Dorogovtsev and J. F. F. Mendes, *Evolution of Networks: From Biological Nets to the Internet and WWW* (Oxford: Oxford University Press, 2003); D. B. West, *Introduction to Graph Theory* (Upper Saddle River, NJ: Prentice-Hall, 1996); S. Wasserman and K. Faust, *Social Network Analysis* (Cambridge: Cambridge University Press, 1994).
8. Edward O. Laumann and David Knoke, "Social Network Theory," in Siegwart Lindenberg, James S. Coleman, and Stefan Nowak, eds., *Approaches to*

Social Theory (New York: Russell Sage Foundation, 1986), 83–104; Duncan J. Watts, "The 'New' Science of Networks," *Annual Review of Sociology* 30 (2004): 243–270; Hannah Knox, Mike Savage, and Penny Harvey, "Social Networks and the Study of Relations: Networks as Method, Metaphor, and Form," *Economy and Society* 35, no. 1 (2006): 113–140.

9. K. S. Cook and J. M. Whitmeyer, "Two Approaches to Social Structure: Exchange Theory and Network Analysis," *Annual Review of Sociology* 18 (1992): 109–127; Ronald S. Burt, "Models of Network Structure," *Annual Review of Sociology* 6 (1980): 79–141; Norman E. Whitten and Alvin W. Wolfe, "Network Analysis," in John J. Honigmann, ed., *Handbook of Social and Cultural Anthropology* (Chicago: Rand McNally, 1973), 717–746.
10. Jean-Guy Vaillancourt, *Papal Power: A Study of Vatican Control over Catholic Lay Groups* (Berkeley: University of California Press, 1980).
11. José Casanova, "Religion, The New Millennium, and Globalization," *Sociology of Religion* 62, no. 4 (2001): 415–441; Timothy A. Byrnes, *Transnational Catholicism in Postcommunist Europe* (Lanham, MD: Rowman and Littlefield, 2001).
12. Tobias Gossling, Leon Oerlemans, and Rob Jansen, eds., *Inside Networks: A Process View on Multi-Organisational Partnerships, Alliances and Networks* (Northampton, MA: Edward Elgar, 2007); Mats Forsgren and Jan Johanson, eds., *Managing Networks in International Business* (Philadelphia: Gordon and Breach, 1992); Bruce McKern, *Managing the Global Network Corporation* (New York: Routledge, 2003).
13. Lawrence S. Welch and ReiJo K. Luostarinen, "Inward-Outward Connections in Internationalization," *Journal of International Marketing* 1, no. 1 (1993): 44–56; Sangmoon Kim and Eui-Hang Shin, "A Longitudinal Analysis of Globalization and Regionalization in International Trade: A Social Network Approach," *Social Forces* 81, no. 2 (2002): 445–471; Heli Konhonen, Reijo Luostarinen, and Lawrence Welch, "Internationalization of SMEs: Inward-Outward Patterns and Government Policy," *Management International Review* 36 (1996): 315–329.
14. Bernard Lewis, *The Crisis of Islam: Holy War and Unholy Terror* (New York: Modern Library, 2003); Fouad Al-Farsy, *Custodian of the Two Holy Mosques: King Fahd Bin Abdul Aziz* (Channel Islands: Knight Communications, 2001).
15. B. Guy Peters and J. Pierre, "Governance Without Government? Rethinking Public Administration," *Journal of Public Administration Research and Theory* 8, no. 2 (1998): 226.
16. R. A. W. Rhodes, *Beyond Westminster and Whitehall* (London: Unwin Hyman, 1988).
17. Mark T. Imperial, "Using Collaboration as a Governance Strategy: Lessons From Six Watershed Management Programs," *Administration and Society* 37, no. 3 (2005): 281–320.
18. Ibid., 308.
19. Several authors have attempted to explain the impact of the globalization process on religions or, conversely, their theological visions of the globalization process, but this is not the route we are following in this analysis. See for example, Richard Falk, *Religion and Humane Global Governance* (New York: Palgrave, 2001); Max L. Stackhouse et al., eds., *God and Globalization*, 3 vols. (Harrisburg, PA: Trinity Press International, 2000); Dwight N. Hopkins, Lois Ann Lorentzen, Eduardo Mendieta, and David Batstone, eds., *Religions/Globalizations: Theories and Cases* (Durham, NC: Duke University Press, 2001).

20. Max Weber, "The Social Psychology of the World Religions," in H. H. Gerth and C. Wright Mills, eds., *From Max Weber: Essays in Sociology*, (London: Routledge and Kegan Paul, 1948), 267–301; Stephen Sharot, *A Comparative Sociology of World Religions* (New York: New York University Press, 2001); Jose Casanova, "Globalizing Catholicism and the Return to a 'Universal' Church," in Susanne H. Rudolph and James Piscatori, eds., *Transnational Religion and Fading States* (Boulder, CO: Westview Press, 1997), 121–143.
21. Pierre Teilhard de Chardin, *La messe sur le monde* (Paris: Editions du Seuil, 1961).
22. Katherine K. Young, "World Religions: A Category in the Making?" in Michel Despland and Gerard Vallee, eds., *Religion in History: The Word, the Idea, the Reality* (Waterloo, IA: Wilfrid Laurier University Press, 1992),111–130; Joachim Wach, *Types of Religious Experience: Christian and Non-Christian* (London: Routledge and Kegan Paul, 1955); Huston Smith, *The Religions of Man* (New York: Harper, 1958); N. J. Demerath, *Crossings the Gods: World Religions and Wordly Politics* (New Brunswick, N.J.: Rutgers University Press, 2001).
23. David Lehmann, "Religion and Globalization," in Linda Woodhead et al., eds., *Religion in the Modern World* (New York: Routledge, 2002), 299–315; see also Kim Knott, *The Location of Religion: A Spatial Analysis* (London: Equinox, 2005), pp. 110–117.
24. Natana J. Delong-Bas, *Wahhabi Islam: From Revival and Reform to Global Jihad* (New York: Oxford University Press, 2004).
25. Irving Hexham and Karla Poewe, *New Religions as Global Cultures* (Boulder, CO: Westview Press, 1997).
26. Mark Juergensmeyer, "Thinking Globally about Religion," in Mark Juergensmeyer, ed., *Global Religions: An Introduction* (New York: Oxford University Press, 2003), 3–13; see also Mark Juergensmeyer, *The Oxford Handbook of Global Religions* (New York: Oxford University Press, 2006).
27. Peter Beyer, *Religions in Global Society* (New York: Routledge, 2006).
28. Ivan Strenski, "The Religion in Globalization," *Journal of the American Academy of Religion* 72, no. 3 (2004): 631–652; Luca Diotallevi, "Roman Catholicism and Religious Globalization," *Antonianum* 79, no. 2 (2004): 345–364.
29. For a literature review of the modern firm seen as a network of nodes, see Joel M. Podolny and Karen L. Page, "Network Forms of Organization," *Annual Review of Sociology* 24 (1998): 57–76.
30. F. Harary, "Graph Theoretic Methods in the Management Sciences," *Management Science* 6, no. 4 (1959): 387–403.
31. John A. Barnes, "Networks and Political Process," in Marc Swartz, ed., *Local-Level Politics*, (Chicago: Aldine); J. Clyde Mitchell, ed., *Social Networks in Urban Situations* (Manchester: Manchester University Press, 1969): 107–130.
32. Peters and Pierre, "Governance without Government?"; W. Powell, "Neither Market Nor Hierarchy: Network Forms of Organization," *Research in Organizational Behavior* 12 (1990): 295–336; Podolny and Page, "Network Forms of Organization"; Grahame Thompson, *Between Hierarchies and Markets: The Logic of Network Forms of Organizations* (Oxford: Oxford University Press, 2003).
33. Starr Roxanne Hiltz and Murray Turoff, *The Network Nation* (Cambridge: MIT Press, 1993); B. Wellman and C. Haythornthwaite, eds., *The Internet in Everyday Life* (Oxford: Blackwell, 2002).

34. Peter R. Monge and Noshir S. Contractor, "Emergence of Communication Networks," in Fredric M. Jablin and Linda L. Putnam, eds., *The New Handbook of Organizational Communication*, (Thousand Oaks, CA: Sage Publications, 2001), 440–502; Bruce Kogut and Gordon Walker, "The Small World of Germany and the Durability of National Networks," *American Sociological Review* 66 (2001): 317–335.
35. Patrick Doreian and Frans N. Stokman, *Evolution of Social Networks* (New York: Gordon and Breach, 1997); D. Gnyawali and R. Madhavan, "Cooperative Networks and Competitive Dynamics: A Structural Embeddedness Perspective," *Academy of Management Review* 26 (2001): 431–445; Barry Wellman and Stephen D. Berkowitz, eds., *Social Structures: A Network Approach* (New York: Cambridge University Press, 1988); J. M. Hite and W. S. Hesterly, "The Evolution of Firm Networks: From Emergence to Early Growth of the Firm," *Strategic Management Journal* 22 (2001): 275–286.
36. James Rosenau and Ernst-Otto Czempiel, eds., *Governance without Government: Order and Change in World Politics* (Cambridge: Cambridge University Press, 1992); Peters and Pierre, "Governance without Government?"
37. Bob Jessop, "Multi-level Governance and Multi-level Metagovernance," in Ian Bache and Matthew Flinders, eds., *Multi-level Governance*, (Oxford: Oxford University Press, 2004).
38. Jan Kooiman, *Governing as Governance* (London: Sage, 2003); Tony Bovaird, "Public Governance: Balancing Stakeholder Power in a Network Society," *International Review of Administrative Sciences* 71, no. 2 (2005): 217–228.
39. Eva Sorensen, "Metagovernance: The Changing Role of Politicians in Processes of Democratic Governance," *American Review of Public Administration* 36, no. 1 (2006): 98–114.
40. Michel Foucault, *Power/Knowledge* (Brighton: Harvester, 1980); Graham Burchell, Colin Gordon, and Peter Miller, eds., *The Foucault Effect: Studies in Governmentality* (Chicago: University of Chicago Press, 1991); Jack Z. Bratich, Jeremy Packer, and Cameron McCarthy, *Foucault, Cultural Studies, and Governmentality* (Albany: State University of New York Press, 2003); Wendy Larner and William Walters, *Global Governmentality: Governing International Spaces* (New York: Routledge, 2004).
41. Janet Fulk, "Global Network Organizations: Emergence and Future Prospects," *Human Relations* 53, no. 1 (2001): 91–99; Janet Newman, "Enter the Transformational Leader: Network Governance and the Micro-politics of Modernization," *Sociology* 39, no. 4 (2005): 717–734; Peter Bogason and Juliet A. Musso, "The Democratic Prospects of Network Governance," *American Review of Public Administration* 36, no. 1 (2006): 3–18; C. Jones, William S. Hesterly, and S. Borgatti, "A General Theory of Network Governance: Exchange Conditions and Social Mechanisms," *Academy of Management Review.* 22, no. 4 (1997): 911–945; Jonathan Davies, "The Hollowing Out of Local Democracy and the 'Fatal Conceit' of Governing without Government," *British Journal of Politics and International Relations* 2, no. 3 (2000): 414–428; Ian Holliday, "Is the British State Hollowing-Out?" *Political Quarterly* 71, no. 2 (2000): 167–176.
42. Arvind Parkhe, Stanley Wasserman, and David A. Ralston, "New Frontiers in Network Theory Development," *Academy of Management Review* 31, no. 3 (2006): 560–568.
43. H. Brinton Milward and Jorg Raab, "Dark Networks: The Structure, Operation, and Performance of International Drug, Terror, and Arms Trafficking Networks." Paper presented at the International Conference on the Empirical Study

of Governance, Management, and Performance, Barcelona, Spain, October 4–5, 2002.

44. Jones, Hesterly, and Borgatti, "A General Theory of Network Governance"; W. J. M. Kickert, E. H. Klinj, and J. F. M. Koppenjan, eds., *Managing Complex Networks: Strategies for the Public Sector* (Thousand Oaks, CA: Sage, 1997); S. H. Park, "Managing an Interorganizational Network: A Framework of the Institutional Mechanism for Network Control," *Organization Studies* 17 (1996): 795–824.
45. Paul F. Diehl, ed., *The Politics of Global Governance: International Organizations in an Interdependent World* (Boulder, CO: Lynne Rienner, 1997); Rosenau and Czempiel, *Governance without Government.*
46. George H. Williams, "John Paul II's Concepts of Church, State, and Society," *Journal of State and Church* 24 (1982): 463–483; Jeff Haynes, "Transnational Religious Actors and International Politics," *Third World Quarterly* 22, no. 2 (2001): 143–158; Jose Casanova, "Global Catholicism and the Politics of Civil Society," *Sociological Inquiry* 66, no. 3 (1996): 356–373.
47. Gerry Stoker, "Public Value Management: A New Narrative for Networked Governance?" *American Review of Public Administration* 36, no. 1 (2006) 41–57; Bovaird, "Public Governance."
48. Gene Burns, *The Frontiers of Catholicism: The Politics of Ideology in a Liberal World* (Berkeley: University of California Press, 1992).
49. Stoker, "Public Value Management." See also Donald F. Kettl, *The Transformation of Governance* (Baltimore: The Johns Hopkins University Press, 2002).
50. Harold K. Jacobson, *Networks of Interdependence: International Organizations and the Global Political System* (New York: Alfred A. Knopf, 1984).
51. Renato Ortiz, "Notes on Religion and Globalization," *Nepantla: Views from South* 4, no 3. (2003): 423–448.
52. Thomas J. Reese, *A Flock of Shepherds: The National Conference of Catholic Bishops* (Kansas City, MO: Sheed and Ward, 1992).
53. P. Monge and N. Contractor, *Theories of Communication Networks* (New York: Oxford University Press, 2003); Lorne L. Dawson and Douglas E. Cowan, eds., *Religion Online: Finding Faith on the Internet* (New York: Routledge, 2004); B. Brasher, *Give Me That Online Religion* (San Francisco: Jossey-Bass, 2001); G. Bunt, *Virtually Islamic: Computer-Mediated Communication and Cyber Islamic Environments* (Cardiff: University of Wales Press, 2000); J. K. Hadden and Douglas E. Cowan, eds., *Religion on the Internet: Research Prospects and Promises* (Greenwich, CT: JAI Press, 2000).
54. Robert Latham and Saskia Sassen, eds., *Digital Formations: IT and New Architectures in the Global Realm* (Princeton, NJ: Princeton University Press, 2005).
55. V. Sambamurthy and Robert W. Zmud, "Arrangements for Information Technology Governance: A Theory of Multiple Contingencies," *MIS Quarterly* 23, no. 2 (1999): 261–290.
56. Dawson and Cowan, eds., *Religion Online*; Hadden and Cowan, eds., *Religion on the Internet*; H. Campbell, "A Review of Religious Computer-Mediated Communication Research," in S. Marriage and J. Mitchell, eds., *Mediating Religion: Conversations in Media, Culture and Religion* (Edinburgh: T. and T. Clark/Continuum, 2003), 213–228; Heidi A. Campbell, *When Religion Meets New Media* (New York: Routledge, 2010).
57. Peter Hebblethwaite, *In the Vatican* (Bethesda, MD: Adler and Adler, 1986).
58. Saskia Sassen, ed., *Deciphering the Global: Its Scales, Spaces, and Subjects* (New York: Routledge, 2007).

59. Joseph Nye and Robert Keohane, "Transnational Relations and World Politics: An Introduction," *International Organization* 25 (1971): 329–346; Andreas Wimmer and Nina Glick Schiller, "Methodological Nationalism, the Social Sciences, and the Study of Migration: An Essay in Historical Epistemology" *Global Networks* 2, no. 4 (2002): 301–334.

NOTES TO CHAPTER 2

1. David Vital, *The Future of Jews* (Cambridge: Harvard University Press, 1990), p. 111.
2. "Land, State, and diaspora in the History of the Jewish Polity," Daniel J. Elazar, accessed November 15, 2010, http://www.jcpa.org/dje/articles2/land-state-polity.htm.
3. Noel M. Tichy, Michael L. Tushman, and Charles Fombrun, "Social Network Analysis for Organizations," *Academy of Management Review* 4, no. 4 (1979): 507–519.
4. Michel Foucault, "Governmentality," in Graham Burchell, Colin Gordon, and Peter Miller, eds., *The Foucault Effect: Studies in Governmentality* (Chicago: University of Chicago Press, 1991), 87–104.
5. Tony Bennett, "Culture and Governmentality," in Jack Z. Bratich, Jeremy Packer, and Cameron McCarthy, eds., *Foucault, Cultural Studies, and Governmentality*, (Albany: State University of New York Press, 2003), 47–63.
6. Michel Foucault, *Power/Knowledge: Selected Interviews and Other Writings, 1972–1977*, ed. Colin Gordon, trans. Colin Gordon, Leo Marshall, John Mepham, and Kate Soper (Brighton: Harvester, 1980).
7. Roland L. Warren, "The Interorganizational Field as a Focus for Investigation," *Administrative Science Quarterly* 12 (1967): 396–419.
8. Leon Oerlemans, Tobias Gossling, and Rob Jansen, "Inside Networks: A Process View on Inter-Organisational Relationships and Networks," in Tobias Gossling, Leon Oerlemans, and Rob Jansen, eds., *Inside Networks: A Process View on Multi-Organisational Partnerships, Alliances and Networks* (Northampton, MA: Edward Elgar, 2007), 3–12; Tobias Gossling, "Inside Relationships: A Review of Institutional Approaches towards Multi-Organisational Partnerships, Alliances and Networks," in *Inside Networks*, 13–44.
9. Harvey E. Goldberg, "Thinking Globally About Judaism," in Mark Juergensmeyer, ed. *The Oxford Handbook of Global Religions* (Oxford: Oxford University Press, 2006), 175–185.
10. Daniel J. Elazar and Stuart A. Cohen, *The Jewish Polity: Jewish Political Organization from Biblical Times to the Present* (Bloomington: Indiana University Press, 1985); S. Ilan Troen, "Introduction: The Post-Holocaust Dynamics of Jewish Centers and Peripheries," in S. Ilan Troen, ed., *Jewish Centers and Peripheries: Europe between American and Israel Fifty Years after World War II* (New Brunswick, NJ: Transaction Publishers, 1999), 1–16.
11. Daniel J. Elazar, *People and Polity: The Organizational Dynamics of World Jewry* (Detroit, MI: Wayne State University Press, 1989).
12. Jerome R. Mintz, *Hasidic People: A Place in the New world* (Cambridge: Harvard University Press, 1992).
13. Michel S. Laguerre, *Global Neighborhoods: Jewish Quarters in Paris, London, and Berlin* (Albany: State University of New York Press, 2008).
14. Jack Wertheimer, *The American Synagogue: A Sanctuary Transformed* (Cambridge: Cambridge University Press, 1987); Jack Wertheimer, ed., *Jews*

in the Center: Conservative Synagogues and their Members (New Brunswick: Rutgers University Press, 2000).
15. Elazar and Cohen, *The Jewish Polity*, 260.
16. Ibid.
17. Ibid., 274.
18. Gilbert S. Rosenthal, ed. *The American Rabbi* (New York: Ktav Publishing House, 1977).
19. Allon Gal, *Envisioning Israel: The Changing Ideals and Images of North American Jews* (Jerusalem: The Magnes Press, and Detroit: Wayne State University Press, 1996), 136.
20. Victor B. Geller, *Orthodoxy Awakens: The Belkin Era and Yeshiva University* (Jerusalem: Urim Publications, 2003), 135.
21. Saul Bernstein, *The Orthodox Union Story: A Centenary Portrayal* (Northvale, NJ: Jason Aronson Inc., 1997), 329.
22. Ibid., 42.
23. Elazar, *People and Polity*, 137.
24. Ibid., 263.
25. Rabbi Israel Goldstein, "The Role of American Rabbis in World Jewish Affairs, Zionism and the State of Israel," in Gilbert S. Rosenthal, ed., *The American Rabbi:* A Tribute on the Occasion of the Bicentennial of the United States, and the Ninety-Fifth Birthday of the New York Board of Rabbis (New York: Ktav Publishing House, 1977), 116.
26. Adam S. Ferziger, *Training American Orthodox Rabbis to Play a Role in Confronting Assimilation* (Ramat Gan, Israel: Bar Ilan University, Faculty of Jewish Studies, 2003), 34.
27. Ibid., 37, 59.
28. Harold Himmelfarb and Sergio Della Pergola, *Jewish Education Worldwide: Crosscultural Perspectives* (Lanham, MD: University Press of America, 1989), 559.
29 Ibid., 560.
30. Elazar, *People and Polity*, 263.
31. Geza Komoroczy, "Jewish Hungary Today: The Jewish Cultural Heritage in the Contemporary Culture of Hungary," in Troen, ed., *Jewish Centers and Peripheries*, 137.
32. Elazar, *People and Polity*, 136.
33. Ibid., 137.
34. Morton H. Narrowe, "From Mosaics to Jews: The Rejection of an Unsuccessful Pattern," in Troen, ed., *Jewish Centers and Peripheries*, 200.
35. David Clayman, "Cooperation and Tensions between American Jewry and Israel over Selected Problems Confronting European Jewry," in Troen, ed., *Jewish Centers and Peripheries*, 370.
36. Sol Encel and Leslie Stein, *Continuity, Commitment, and Survival: Jewish Communities in the Diaspora* (London: Praeger, 2003), 37.
37. Marlena Schmool, "British Jewry: Prospects and Problems," in Troen, ed., *Jewish Centers and Peripheries*, 240.
38. Daniel J. Elazar and Rela Mintz Geffen, *The Conservative Movement in Judaism* (Albany: State University of New York Press, 2000), 129–130.
39. Encel and Stein, *Continuity, Commitment, and Survival*, 169.
40. Elazar, *People and Polity*.
41. Troen, "Introduction," in Troen, ed., *Jewish Centers and Peripheries*, 16.
42. Elazar, *People and Polity*, 70.
43. Clayman, "Cooperation and Tensions," in Troen, ed., *Jewish Centers and Peripheries*, 370.
44. Ibid., 149.

45. See, for example, Manfred Gerstenfeld and Avraham Wyler, "Technology and Jewish Life," *Jewish Political Studies Review* 18, nos. 1–2 (2006): 1–22; Robin Treistman, "Judaic Law in an Internet World," available online at http://www.ifla.org/IV/ifla66/papers/020-160e.htm, last accessed June 22, 2010.
46. Menachem Friedman, *The Haredi (Ultra-Orthodox) Society: Sources, Trends and Processes* (in Hebrew) (Jerusalem: Jerusalem Institute for Israel Studies, 1991),149.
47. Mike Kamber, "Ban the Web? Not Lubavitch Jews," *Wired*, January 19, 2000.
48. Allon Gal and Alfred Gottschalk, eds., *Beyond Survival and Philanthropy: American Jewry and Israel* (Cincinnati, OH: Hebrew Union College, 2000); David Mittelberg, *The Israel Connection and American Jews* (London: Praeger, 1999); Richard G. Hirsch, *From the Hill to the Mount* (Jerusalem: Gefen Publishing House, 2000).
49. Michele Chabin, "U.S. Orthodox Convert Denied Israel Marriage by Chief Rabbi," *Jewish Bulletin of Northern California* (1997), available online at http://www.jewishsf.com/content/2-0-/module/displaystory/ story_id/6943/ edition_id/131/format/html/displaystory.html, last accessed June 22, 2010.
50. The focus on networks must also pay attention to two other aspects of the process: subnetwork-subnetwork relations (friction, tension, competition, cooperation, disappearance, appearance) and the subnetwork as a node.

NOTES TO CHAPTER 3

The epigraph is from S. P. Colbi, "The Christian Establishment in Jerusalem," in Joel L. Kraemer, ed., *Jerusalem: Problems and Prospects* (New York: Praeger Publishers, 1980), 153.

1. Thomas J. Reese, *Inside the Vatican: The Politics of Organization of the Catholic Church* (Cambridge: Harvard University Press, 1996).
2. Corrado Pallenberg, *Inside the Vatican* (New York: Hauthorne, 1960); George Seldes, *The Vatican: Yesterday, Today, Tomorrow* (New York: Harper, 1934); Jean Neuvecelle, *The Vatican: Its Organization, Customs, and Way of Life* (New York: Criterion, 1955).
3. Benedict Williamson, *The Treaty of the Lateran* (London: Burns Oates & Washbourne, 1929).
4. Several studies of the globalization of the Catholic Church are now available. See, for example, Jose Casanova, *Public Religions in the Modern World* (Chicago: University of Chicago Press, 1994); Ivan Vallier, "The Roman Catholic Church: A Transnational Actor," in Robert O. Keohane and Joseph Nye, eds., *Transnational Relations and World Politics* (Cambridge: Harvard University Press, 1972): 479–502; Eric O. Hanson, *The Catholic Church in World Politics* (Princeton, NJ: Princeton University Press, 1987); Jean-Guy Vaillancourt, *Papal Power: A Study of Vatican Control over Catholic Lay Groups* (Berkeley: University of California Press, 1980); Ralph Della Cava, "Vatican Policy, 1978–1990: An Updated Overview," *Social Research* 59, no. 1 (1992): 169–199; Ralph Della Cava, "Financing the Faith: The Roman Catholicism," *Journal of Church and State* 35, no. 1 (1993): 1101–1122; Penny Lernoux, *People of God: The Struggle for World Catholicism* (New York: Viking, 1989), and others.
5. Yasmin Abdullah, "The Holy See at United Nations Conferences: State or Church?" *Columbia Law Review* 96, no. 7 (1996): 1835–1875; "The 1929 Lateran Treaty," in John W. Wheeler-Bennett, ed.,*Documents on International*

Affairs (London: Oxford University Press, 1930), 216–241; Hyginus Eugene Cardinale, *The Holy See and the International Order* (Gerrards Cross, England: Colin Smythe, 1976).

6. Pierre Blet S.J., *Histoire de la répresentation diplomatique du Saint Siége: Des origines á l'aube du XIXésiécle* (Vatican City: Archivo Vaticano, 1990); Henri Biaudet, *Les nonciatures apostoliques permanentes jusqu'en 1648* (Helsinki: Suomalainen Tiedeakatemia, 1910); P. Richard, "Les origines des nonciatures permanentes," *Revue d'Histoire Ecclésiastique* 7 (1906): 52–70, 317–338; Cuthbert H. Turner, *Ecclesiae occidentalis monumenta juris antiquissima: Canonum et conciliorum Graecorum interpret*, 5 vols. (Oxford: Clarendon, 1899–1933).
7. Walter Powell, "Neither Market Nor Hierarchy: Network Forms of Organization,"*Research In Organizational Behavior* 12 (1990): 295–336.
8. Alan Wells and Mary Lou De Ciantis, "World Systems: Transnational Organizations and the Catholic Church."*Free Inquiry in Creative Sociology* 13,no. 1 (1985): 70–74.
9. Sumantra Ghoshal and Christopher A. Bartlett, "The Multinational Corporation as an Interorganizational Network," *Academy of Management Review* 15 (1990): 603–625.
10. Joseph Bernhart, *The Vatican as a World Power* (London: Longmans, Green and Co., 1939); Cardinale, *The Holy See and International Order*; Carl C. Eckhardt, *The Papacy and World Affairs* (Chicago: University of Chicago Press, 1937); Paul Hoffman, *O Vatican! A Slightly Wicked View of the Holy See* (New York: Congdon and Weed, 1984); Francis X. Murphy, "Vatican Politics: Structure and Function,"*World Politics* 26, no 4 (1974): 542–559.
11. Rachel Mary Harrington, "The Applicability of the Principle of Subsidiarity According to the 'Code of Canon Law,'" Ph.D. dissertation, University of Ottawa, Canada, 1997.
12. Francisco Javier Urritia, Delegation of the Executive Power of Governance. *Studia Canonica. Revue Canadienne de Droit Canonique* 19, no.1 (1985): 339–355.
13. Renee de la Torre Castellanos, "The Catholic Diocese: A Transversalized Institution." *Journal of Contemporary Religion* 17, no.3 (2002): 303–316.
14. T. Hefferan, *Twinning Faith and Development: Catholic Parish Partnering in the US and Haiti* (Bloomfield, CT: Kumarian Press, 2007); see also Helen Rose Ebaugh and Janet Saltzman Chafetz, eds.,*Religion and the New Immigrants: Continuities and Adaptations in Immigrant Congregations* (Walnut Creek: AltaMira Press, 2000); Manuel A.Vasquez and Marie F. Marquardt, *Globalizing the Sacred: Religion Across the Americas* (New Brunswick, NJ: Rutgers University Press, 2003); Helen Rose Ebaugh and Janet Saltzman Chafetz, eds.,*Religion Across Borders: Transnational Immigrant Networks* (Walnut Creek, CA: AltaMira Press, 2002).
15. Michel S. Laguerre, "The Transglobal Network Nation: Diaspora, Homeland and Hostland,"in Eliezer Ben-Rafael and Yitzhak Sternberg, eds., *Transnationalism: Diasporas and the Advent of a New (Dis) Order* (Boston: Brill, 2009), 195–210.
16. Catholic Church, *Code de Droit Canonique Bilingue et Annoté* (Montreal: Wilson & Lafleur, 1999).
17. For a more elaborate and comprehensive discussion of the ecclesiastical circumscriptions of the Catholic Church, see Juan Ignacio Arrieta, *Governance Structures Within the Catholic Church* (Montreal: Wilson and Lafleur, 2000), 169–193.
18. J. B. D'Onorio, *Le pape et le gouvernement de l'Église* (Paris: Fleurus Tardy, 1992).

19. *Code De Droit Canonique Bilingueet Annoté.*
20. John Paul II, *Norms for the Distribution of Priests* (Vatican City, 1980); John Paul II, *Redemptoris Missio* (Vatican City, 1990); Congregation for the Evangelization of Peoples, *Norms Governing Diocesan Priests Serving Abroad* (Vatican City, 2001); Congregation for the Evangelization of Peoples, *Instruction on the Sending Abroad and Sojourn of Diocesan Priests from Mission Territories* (Vatican City, April 25, 2001).
21. Peggy Levitt, *God Needs No Passport: Immigrants and the Changing American Religious Landscape* (New York: The New Press, 2007).
22. William L. Smith, *Irish Priests in the US: A Vanishing Subculture* (Dallas: University Press of America, 2004).
23. Dean R. Hoge, *The Future of Catholic Leadership: Responses to the Priest Shortage* (Kansas City, MO: Sheed and Ward, 1987); SarithaRai, "Short on Priests: U.S. Catholics Outsource Prayers to Indian Clergy," *New York Times*, June 13, 2004; Steven G. Vegh, "Foreign Priests Fill Persistent Vacancies in the U.S. Dioceses: Dozens Have Come to VA from Abroad," *The Virginia Pilot and the Ledger-Star*, January 4, 2004, A1.
24. Dean R. Hoge and Aniedi Okure, *International Priests in America: Challenges and Opportunities* (Collegeville, MN: Liturgical Press, 2006).
25. Since Vatican II, there has been an incremental process of internationalization of the Curia. See Cava, "Vatican Policy, 1978–90"; anonymous "Bilan du Pontificat, La Réforme de la Curie," *Informations Catholiques Internationales*, no. 530 (September 1978), 31–32; Paul VI, *Regimini Ecclesiae Universalis* (August 1967 decree).
26. For analyses of various aspects of the finances of the Vatican, see John F. Pollard, *Money and the Rise of the Modern Papacy: Financing the Vatican, 1850–1950* (New York: Cambridge University Press, 2005); Alain Buzelais, "De pecunia vaticani: Ou en sont les finances du Vatican," *Reperes: Bulletin Economiqueet Financier de la Banque Internationale a Luxembourg (BIL)*47 (1996).
27. Willard M. Oliver, "500 Years of Service to the Popes,"*Crime and Justice International* 22, no. 90 (2006): 4–10.
28. http://www.cnewa.org/generalpg-verus.aspx?pageID=48; last accessed on July 12, 2010.
29. Michael J. L. La Civita, "After the Storm,"*ONE Magazine* 29, no.3 (May 2003); Peter Lemieux, "A Flicker of Candlelight Amid the Darkness,"*ONE Magazine* 25, no. 5 (September 2003).
30. http://www.cnewa.org/generalpg-verus.aspx?pageID=48; last accessed on July 12, 2010.
31. For analyses of the efficient use of the mass media by the Vatican, see Omar Calabrese, "Religious Communication: A Little Structural Anthropology is Necessary," *Religioni e Societa* 20, no. 53 (2005): 63–67; Jerome Montes, "Vatican Diplomacy at the Start of the New Millennium," *Futuribles* 284 (March 2003): 45–53; Roberto J. Blancarte, "Religion, Mass Media, and Power," *Religioni e Societa* 13, no. 30 (1998): 22–34; and Italo Moscati, "A Taboo: Religion and the Mass Media," *Religioni e Societa* 13, no. 30 (1998): 69–77.
32. Richard Owen, "Vatican Cashes in by Putting Price on the Pope's Copyright," *The Times* (London), January 23, 2006.
33. International Telephone and Telegraph Corporation. *International Communications Review* 7, no.1 (January 1931).
34. Ralph Della Cava, "Religious Resource Networks: Roman Catholic Philanthropy in Central and East Europe," in Susanne Hoeber Rudolph and James Piscatori, eds.,*Transnational Religion: Fading States* (Boulder, CO: Westview Press, 1997),113–211.

35. Shawn Tully and Marta F. Dorion, "The Vatican's Finances," *Fortune* 116, no. 14 (December 21, 1987), available online at http://money.cnn.com/magazines/fortune/fortune_archive/1987/12/21/70001/index.htm,last accessed November 9, 2008.
36. Steve Hamm, "O Click All Ye Faithful: The Nun Who Launched the Vatican's Web Site is at Work on a MySpace for Catholics," *Business Week,*May 8, 2006.
37. John Paul II, "Internet: A New Forum for Proclaiming the Gospel. Message of the Holy Father for the 36th World Communication Day," *Pontificium Consilium De Communicationibus Socialibus* May 12, 2002. http://www.vatican.va/roman_curia/pontifical_councils/pccs/documents/rc_pc_pccs_doc_20020228_church-internet_en.html; last accessed on July 12, 2010.
38. John-Paul II, "Internet: A New Forum for Proclaiming the Gospel. Message of the Holy Father for the 36th World Communications Day," *Pontificium-Consilium De Communicationibus Socialibus*, May 12, 2002. Other documents issued by the Pontifical Council for Social Communications include "The Church and Internet" and "Ethics and Internet"
39. Nicola Clark, "The Expanding Vision of Vatican's Holy See," *International Herald Tribune*, Retrieved June 6, 2010, http://www.iht.com/articles/2004/04/26/wireless26_ed3_.php.
40. http://www.vatican.va/news_services/television/index_it.htm; last accessed on July 12, 2010. Centro Televisivo Vaticano-Index. "Servizi di Informazione."
41. Cindy Wooden, "Pope Asks Young Catholics to Use Technology to Share their Faith,"*Catholic News Service*, January 23, 2009.
42. James Fredericks, "The Catholic Church and the Other Religious Paths: Rejecting Nothing that is True and Holy," *Theological Studies* 64, no.2 (2003): 225–254.
43. Henry T. King, Jr., and Marc A. Le Forestier, "Papal Arbitration,"*Dispute Resolution Journal* 52, no. 3 (1997): 74–80.
44. Catalina Romero speaks of the dual role of the church in respect to its global dimension. She states that "on one side, the church functions as a religious institution within a central government structure which makes global decisions . . . within a hierarchical political regime. On the other side, global networks of Catholics within local churches . . . function at the level of everyday life." Catalina Romero, "Globalization, Civil Society and Religion from a Latin American Standpoint," *Sociology of Religion* 62, no. 4 (2001): 475.
45. Roger Finke and Patricia Wittberg, "Organizational Revival From Within: Explaining Revivalism and Reform in the Roman Catholic Church," *Journal for the Scientific Study of Religion* 39, no. 2 (2000): 154–170.
46. Romero, "Globalization, Civil Society and Religion," 475.
47. Andrew M. Greeley, "The Revolutionary Event of Vatican II: How Everything Changed," *Commonwealth* 25, no. 15 (1998): 14–20. Vatican II recalibrated the structure of authority or governance of the Catholic Church. The council elevated the status of the bishops (one among equals in reference to the status of the pope) and of the laity (they are a ministerial people and this function cannot be attributed exclusively to priests) and downgraded the status of the pope (first among equals in reference to the status of the bishops) and the priests (not better or higher than the laity). Network governance must be seen within the context of this new network architecture.
48. Mark R. Kowalewski, "Firmness and Accommodation: Impression Management in Institutional Roman Catholicism," *Sociology of Religion* 54, no. 2 (Summer 1993): 207; Avery Dulles, *Models of the Church* (New York: Doubleday, 1978).

49. Gregory Baum, "Le Rapport Dumont: Democratiser l'Église Catholique," *Sociologie et Sociétés* 22, no. 2 (1990): 115–125.

NOTES TO CHAPTER 4

1. John L. Esposito, ed., *The Oxford Encyclopedia of the Modern Islamic World* (Oxford: Oxford University Press, 1995); Esposito, ed., *Voices of Resurgent Islam* (Oxford: Oxford University Press, 1983); Esposito, *Islam: The Straight Path* (Oxford: Oxford University Press, 1988); Sophie Gilliat-Ray, "Invitation and Enculturation: The Dissemination and Development of Islam in Contemporary Societies," in John Wolffe, ed., *Global Religious Movements in Regional Context* (Aldershot, UK: Ashgate, 2002), 157–194; Miriam Cooke and Bruce B. Lawrence, eds., *Muslim Networks From Hajj to Hip Hop* (Chapel Hill: University of North Carolina Press, 2005); Peter Mandaville, *Transnational Muslim Politics: Reimagining the Umma* (New York: Routledge, 2001); Stefano Allievi and Jorgen S. Nielsen, eds., *Muslim Networks and Transnational Communities in and across Europe* (Leiden: Brill, 2003); Yvonne Yazbeck Haddad, "The Globalization of Islam," in John L. Esposito, ed., *The Oxford History of Islam* (Oxford: Oxford University Press, 1999), 601–641.
2. George Snavely Rentz, Jr., "Muhammad ibn 'Abd al-Wahhab (1703/4–1792) and the Beginnings of the Unitarian Empire in Arabia," Ph.D. dissertation, University of California, Berkeley, 1948; Henri Laoust, "Ibn Abd al-Wahhab, Muhammad," in B. Lewis, V. C. Menage, Ch. Pellat, and J. Schacht, eds., *The Encyclopedia of Islam*, vol. 3 (Leiden: Brill, 1971); Natana J. DeLong-Bas, *Wahhabi Islam: From Revival and Reform to Global Jihad* (Oxford: Oxford University Press, 2004); Ayman al-Yassini, "Ibn 'Abd al-Wahhab, Muhammad," in *The Oxford Encyclopedia of the Modern Islamic World*; John O. Voll, "Muhammad Hayat al-Sindi and Muhammad Ibn 'Abd al-Wahhab: An Analysis of an Intellectual Group in Eighteenth Century Medina," *Bulletin of the School of Oriental and African Studies* 38, no. 1 (1975): 32–39; Voll, *Islam: Continuity and Change in the Modern World* (Syracuse, NY: Syracuse University Press, 1994); Abd Allah al-Salih al-Uthaymin, "Muhammad Ibn 'Abd al-Wahhab: The Man and His Works," Ph.D. dissertation, Edinburgh University, 1972; Michael Cook, "On the Origins of Wahhabism," *Journal of the Royal Asiatic Society of Great Britain and Ireland* 2, no. 2 (1992): 191–202; Hamid Algar, *Wahhabism: A Critical Essay* . Oneonta, NY: Islamic Publications International, 2002.
3. Georges De Bouteiller, *L'Arabie Saoudite: Cité de dieu, cité des affaires, puissance internationale* (Paris: Presses Universitaires de France, 1981), 7.
4. Quoted in Fouad al-Farsy, *Modernity and Tradition: The Saudi Equation* (London: Kegan Paul International, 1990), 20–21.
5. Ibid., 37.
6. Mamoun Fandy, *Saudi Arabia and the Politics of Dissent* (New York: Palgrave, 1999); Christine Moss Helms, *The Cohesion of Saudi Arabia: Evolution of Political Identity* (Baltimore: The Johns Hopkins University Press, 1981); Madawi al-Rasheed, *A History of Saudi Arabia* (New York: Cambridge University Press, 2002); Madawi Al-Rasheed and Robert Vitalis, eds., *Counter-Narratives: History, Contemporaary Society, and Politics in Saudi Arabia and Yemen* (New York: Palgrave Macmillan, 2004); Ayman Al-Yassini, *Religion and State in the Kingdom of Saudi Arabia* (Boulder, CO: Westview Press, 1985).

7. Antoine Basbous, *L'Arabie Saoudite en question: Du Wahhabisme á Bin Laden, aux Origines de la Tourmente* (Paris: Perrin, 2002), 123; Joseph A. Kechichian, "The Role of the Ulama in the Politics of an Islamic State: The Case of Saudi Arabia," *International Journal of Middle Eastern Studies* 18 (1986).
8. Kechichian, "The Role of the Ulama," 53–71.
9. Cary Fraser, "In Defense of Allah's Realm: Religion and Statecraft in Saudi Foreign Policy Strategy," in Susanne H. Rudolph and James Piscatori, eds., *Transnational Religion and Fading States*, (Boulder, CO: Westview Press, 1997).
10. Sam S. Souryal, "The Religionization of a Society: The Continuing Application of Shariah Law in Saudi Arabia," *Journal for the Scientific Study of Religion* 26, no. 4 (1987): 429–449.
11. For an elaboration on the history of power, authority, and governance in the evolution of Islam, see Hamid Dabashi, *Authority in Islam: From the Rise of Muhammad to the Establishment of the Ummayads* (New Brunswick, NJ: Transaction Books, 1989).
12. Raymond E. Miles, Charles C. Snow, John A. Mathews, Grant Miles, and Henry J. Coleman, Jr., "Organizing in the Knowledge Age: Anticipating the Cellular Form," *Academy of Management Executive* 11, no. 4 (1997): 7–20. See also Charles C. Snow, Raymond E. Miles, and Brent B. Allred, "Managing Leading-Edge Multinational Corporations," in Bruce McKern, ed., *Managing the Global Network Corporation* (New York: Routledge, 2003), 201–219.
13. Delong-Bas, *Wahhabi Islam*, 291.
14. "1,947,578 Copies of Quran Distributed within the Kingdom and Abroad," available online at http://www.saudinf.com/main/y7578.htm (last accessed June 22, 2010).
15. Anonymous, "Blessings and Success for the Islamic World," *Saudi Arabia* 9, no. 2 (Summer 1992): 20–23.
16. See "WAMY Builds Mosques," available online at http://www.saudinf.com/main/y8450.htm (last accessed June 22, 2010).
17. "WAMY Builds Mosques in Southern Asia," available online at http://www.saudinf.com/main/y6370.htm (last accessed November 11, 2008).
18. "Prince Sultan Donates $2 million for Kazakhstan Mosque," available online at http://www.saudinf.com/main/y7544.htm (last accessed June 22, 2010).
19. "New WAMY Secretary-General Appointed," available online at http://saudiembassy.net/2002News/News/Isldetail.asp?cIndex=519 (last accessed June 22, 2010).
20. James Wynbrandt, *A Brief History of Saudi Arabia* (New York: Facts On File, 2004), 238.
21. The Agency of Religious Affairs (in arabic, translated by Manar Hassan), available online at http://www.al-islam.com/arb/Minsteryinfo.asp?f=Mister00011.htm (last accessed November 11, 2008).
22. "Islamic Projects Sponsored by Saudi Arabia," available online at http://saudiembassy.net/archive/1996/news/page100.aspx (last accessed November 11, 2008).
23. "Report on Islamic Centers Worldwide," available online at http://saudiembassy.net/2000News/News/Isldetail.asp?cIndex=2238 (last accessed November 11, 2008).
24. "Saudi Schools Overseas," available online at http://saudiembassy.net/about/country-information/education/Saudi_schools_overseas.aspx (last accessed November 11, 2008).
25. "Prince Bandar Inaugurates Islamic Center in Ohio," available online at http://saudiembassy.net/archive/1995/news/page17.aspx (last accessed November 11, 2008).

26. Anonymous. "Serving the Global Islamic Community," *Saudi Arabia* 15, no. 3 (Fall 1998): 10.
27. Basbous, *L'Arabie Saoudite en question*, 147.
28. Anonymous, "Aiding the Needy across the World," *Saudi Arabia* 11, no. 4
29. http://iu.edu.sa/web/content.aspx?id-360 (accessed November 17, 2010) (Winter 1995): 12–17.
30. Basbous, *L'Arabie Saoudite en question*, 155.
31. Al-Yassini, *Religion and State in the Kingdom of Saudi Arabia* , 73.
32. Several studies on the hajj are available. See for example, Muhammad Asad, *The Road to Mecca* (London: Max Reinhardt, 1954); 'Abd al-'Aziz bin Abdullah bin Baz, *A Guide to Hajj, Umrah and Visitng the Prophet's Mosque* (Riyadh: IFTA, and Propagation, 1990); Sir Richard F. Burton, *Personal Narrative of a Pilgrimage to Al-Madinah and Meccah*, 2 vols. (New York: Dover, 1964); Ezzedine Gellouz, *Mecca: The Muslim Pilgrimage* (New York: Paddington Press, 1979); Hamza Kaidi, *Mecca and Medinah Today* (Paris: Les Éditions J.A., 1980); Z. I. Khan and Y. Zaki, eds. *Hajj in Focus* (London: Open Press, 1986); Hussein Yoshi Hirashima, *The Road to Holy Mecca* (Tokyo: Kodansha International, 1972); Sardar Ziauddin and M. A. Zaki Badawi, *Hajj Studies*, vol. 1 (London: Croom Helm, 1977); Michael N. Pearson, *Pilgrimage to Mecca: The Indian Experience 1500–1800* (Princeton, NJ: Markus Wiener, 1996); Ahmad Kamal, *The Sacred Journey: Being Pilgrimage to Makkah* (New York: Duell, Sloane, and Pearce, 1961); Saida Miller Khalifa, *The Fifth Pillar: The Story of a Pilgrimage to Mecca and Medina* (Hicksville, NY: Exposition Press, 1977); Muhammad Zafrulla Khan, *Pilgrimage to the House of Allah* (London: London Mosque, 1968); Russel King, "The Pilgrimage to Mecca: Some Geographical and Historical Aspects," Erdkunde 26 (1972): 61–73; David E. Long, *The Hajj Today: A Survey of the Contemporary Makkah Pilgrimage* (Albany: State University of New York Press, 1979); G. A. W. Makky, *Mecca, the Pilgrimage City: A Study of Pilgrim Accommodation* (London: Croom Helm, 1978); I. M. Shair and P. P. Karan, "Geography of the Islamic Pilgrimage, *GeoJournal* 3, no. 6 (1979): 599–608; Adu Bakr Siraj ad-Din, "Pilgrimage to Mecca," *Studies in Comparative Religion* 1 (1967): 171–181; Ziauddin Sardar, "The Future of Hajj: Some Basic Considerations," *Islamic Culture* 58, no. 4 (1984): 307–326; Gwyn Rowley and S. A. S. El-Hamdan, "The Pilgrimage to Mecca: An Exploratory and Predictive Model," *Environment and Planning A* 10 (1978): 1053–1071; Gwyn Rowley and S. A. S. El-Hamdan, "Once a Year in Mecca," *Geographical Magazine* 49 (1977): 753–759; William R. Roff, "Sanitation and Security: The Imperial Powers and the Nineteenth-Century Hajj," *Arabian Studies* 6 (1982): 143–160; Francis E. Peters, *The Hajj: The Muslim Pilgrimage to Mecca and the Holy Places* (Princeton: Princeton University Press, 1994), and others.
33. See http://saudiembassy.net/2002news/news/hajdetail.asp?cIndex=565 (last accessed November 11, 2008).
34. "Hajj Minister Talks about New Umrah Rules," http://saudiembassy.net/archive/2001/news/page307.aspx (last accessed November 11, 2008).
35. "Minister Gives Assurances of Health Preparations for 1424 Hajj," available online at http://saudiembassy.net/archive/200/news/page796.aspx (last accessed November 11, 2008).
36. "King Fahd to Host 1,500 Pilgrims from Asia and Europe," available online at http://saudiembassy.net/2003/news/page551.aspx (last accessed November 11, 2008); "King Fahd Hosts More Pilgrims," available online at

http://saudiembassy.net/2003News/News/Hajdetail.asp?cIndex=561 (last accessed November 11, 2008).

37. "Saudi Arabian Airline to Transport 100,000 Iranian Pilgrims this Hajj Season," available online at http://saudiembassy.net/2005/news/page887.aspx (last accessed November 11, 2008).
38. "Over 1.2 Million Hajj Visas Issued," available online at http://saudiembassy.net/200/news/page859.aspx (last accessed November 11, 2008).
39. "Visas for 1426 Hajj," available online at http://www.saudiembassy.net/Travel/Hajj.asp (last accessed November 11, 2008).
40. David E. Long, *The Kingdom of Saudi Arabia* (Gainesville: University Press of Florida, 1997), 102.
41. "Hajj 1420 Sacrificial Meat Being Distributed," available online at http://saudiembassy.net/2000News/News/Hajdetail.asp?cIndex=626 (last accessed November 11, 2008).
42. New brands of soft drinks (Mecca-Cola, Zamzam-Cola, Arab-Cola, and Qibla-Cola) are globally commercialized to sustain and reinforce the identity of Muslin dietary practices. See Patrick Heanni, *L'Islam de marché: L'autre revolution conservatrice* (Paris: Seuil, 2005), 55.
43. "Islamic University of Medinah: Shining Example of Islamic Education," *Saudi Arabia: The Magazine* 16 (4) available online at http://saudiembassy.net/Publications/MagWinter00/Madinah.htm (last accessed November 11, 2008).
44. "Education Ministry to Send Teachers Overseas to Work," available online at http://saudiembassy.net/200/news/page290.aspx (last accessed November 11, 2008).
45. Afzal Khan, "American Muslims Perform Hajj, Celebrate Eid-ul Adha," available online at http://www.america.gov/st/washfile-english/2005/January/20050121173803cpataruk0.6693384.html (last accessed June 22, 2010).
46. Muhammad Abdullah al-Mani and Abd ur-Rahman Sbit as-Sbit, *Cultural Policy in the Kingdom of Saudi Arabia* (Paris: The UNESCO Press, 1981), 60.
47. Douglas A. Boyd, "Saudi Arabian Television," *Journal of Broadcasting* 15, no. 1 (1970–1971): 73–78; William A. Rugh, "Saudi Mass Media and Society in the Faisal Era," in Willard A. Beling, ed., *King Faisal and the Modernisation of Saudi Arabia* (London: Croom Helm, 1980).
48. Nasser Ibrahim Rashid and Esber Ibrahim Shaheen, *Saudi Arabia All You Need to Know* (Joplin, MO: International Institute of Technology, 1995), 141.
49. Fouad Al-Farsy, *Custodian of the Two Holy Mosques King Fahd Bin Abdul Aziz* (Channel Islands: Knight Communications, 2001), 33.
50. "New Internet Site on Hajj to Be Launched Saturday," available online at http://saudiembassy.net/2001/news/page522.aspx (last accessed November 11, 2008).
51. See for example, Bruce B. Lawrence, "Allah On-Line: The Practice of Global Islam in the Information Age," in Stewart M. Hoover and Lynn Schofield Clark, eds., *Practicing Religion in the Age of the Media: Explorations in Media, Religion, and Culture*, (New York: Columbia University Press, 2002), 236–253; G. Bunt, *Virtually Islamic: Computer-Mediated Communication and Cyber Islamic Environments* (Cardiff: University of Wales Press, 2000). Sam Cherribi, *The Growing Islamizaion of Europe*. In John L. Esposito and François Burgat, Modernizing Islam: Religion in the Public Sphere in Europe and the Middle East. New Brunswick, NJ: Rutgers University Press, 2003, pp. 192–213.
52. Basbous, *L'Arabie Saoudite en question*, 156.

53. William B. Quandt, *Saudi Arabia in the 1980s: Foreign Policy, Security, and Oil* (Washington, D.C.: The Brookings Institution, 1981), 37.
54. Long, *The Kingdom of Saudi Arabia*, 98.
55. Alexei Vassiliev, *The History of Saudi Arabia* (London: Saqi Books, 1998), 396, 471.
56. Middle East Media Research Institute, University of Pennsylvania, "Saudi Arabia," in *MEMRI Report* (London: Croom Helm, 1985), 18.
57. Basbous, *L'Arabie Saoudite en question*, 81.

NOTES TO CHAPTER 5

1. The situation of the global religion has many similarities with the case of the multinational firm, as can be seen in John H. Dunning and Jean-Louis Mucchielli, eds., *Multinational Firms: The Global-Local Dilemma* (London: Routledge, 2002).
2. Richard Mead, *International Management: Cross-Cultural Dimensions* (Malden, MA: Blackwell, 1998).
3. David Coghlan, "Corporate Strategy in Catholic Religious Orders," *Long Range Planning* 20, no. 1 (1987): 44–51.
4. Arvind V. Phatak, Rabi S. Bhagat, and Roger J. Kashlak, *International Management: Managing in a Diverse and Dynamic Global Environment* (New York: McGraw-Hill, 2005).
5. Mats Forsgren and Jan Johanson, eds., *Managing Networks in International Business* (New York: Gordon and Breach, 1992).
6. Barbara Czarniawska, ed., *Organization Theory: Current Trends and Disciplinary Reflection* (Northampton, MA: Edward Elgar, 2006).
7. Tobias Gossling, Leon Oerlemans, and Rob Jansen, eds., *Inside Networks: A Process View on Multi-Organizational Partnerships, Alliances and Networks* (Northampton, MA: Edward Elgar, 2007).
8. Jean-René Bertrand, "Religions et territoires," in Jean-René Bertrand and Collette Muller, eds., *Religions et Territoires* (Paris: L'Harmattan, 1999), 11–16; see also R. D. Sack, *Human Territoriality, Its Theory and History* (Cambridge: Cambridge University Press, 1986).
9. Jacques Palard, *Le gouvernement de l'Église Catholique: Synodes et exercices du pouvoir* (Paris: Cerf, 1997).
10. Joseph G. Jabbra and O. P. Dwivedi, eds., *Administrative Culture in a Global Context* (Whitby, Ontario: de Sitter Publications, 2005); Piero Morosini, *Managing Cultural Differences: Effective Strategy and Execution across Cultures in Global Corporate Alliances* (New York: Pergamon, 1998).
11. Madeleine Adriance, "The Paradox of Institutionalization: The Roman Catholic Church in Chile and Brazil," *Sociological Analysis* 53 (1992): 51–62; Frank Lewins, "Continuity and Change in a Religious Organization: Some Aspects of the Australian Catholic Church," *Journal for the Scientific Study of Religion* 16, no. 4 (1977): 371–382; Lorenzo Livieres Banks, "Notes Toward Understanding the Socio-Historic Functions of the Apostolic Roman Catholic Church of Paraguay," *Revista Paraguaya de Sociologia* 7, no. 17(1970): 123–132.
12. Robert T. Moran, Philip R. Harris, and William G. Stripp, *Developing the Global Organization: Strategies for Human Resource Professionals* (Houston: Gulf , 1993).
13. Fiona Moore, *Transnational Business Cultures: Life and Work in a Multinational Corporation* (London: Ashgate, 2005).

14. Much more can be done in this area to facilitate the global management of these religions. For new developments in this area concerning global firms, see Robert J. Mockler and Dorothy G. Dologite, *Multinational Cross-Cultural Management: An Integrative Context-Specific Process* (Westport, CT: Quorum Books, 1997).
15. Michael A. Zigarelli, "Catholic Social Teaching and the Employment Relationship: A Model for Managing Human Resources in Accordance with Vatican Doctrine," *Journal of Business Ethics* 12, no. 1 (1993): 75–82.
16. Laurie Goodstein, "Pope Adds Meetings with Jewish Leaders, Including Synagogue Visit, to U.S. Itinerary," *New York Times*, April 4, 2008, A12.

Bibliography

Abdullah, Yasmin. 1996. The Holy See at United Nations Conferences: State or Church? *Columbia Law Review* 96 (7): 1835–1875.

AbuKhalil, As'ad. 2004. *The Battle for Saudi Arabia: Royalty, Fundamentalism, and Global Power.* New York: Seven Stories.

Adriance, Madeleine. 1992. The Paradox of Institutionalization: The Roman Catholic Church in Chile and Brazil. *Sociological Analysis* 53: 51–62

Agency of Religious Affairs. (translated from Arabic by Manar Hassan) (http://www.al.islam.com/arb/Minsteryinfo.asp?f=Mister00011.htm)

Allievi, Stefano and Jorgen S. Nielsen. 2003. *Muslim Networks and Transnational Communities in and across Europe.* Leiden: Brill.

Anonymous. 1978. Bilan du Pontificat: La Reforme de la Curie. *Informations Catholiques Internationales* 530: 31–32.

———. 1992. Blessings and Success for the Islamic World. *Saudi Arabia* 9 (2): 20–23.

———. 1998. Serving the Global Islamic Community. *Saudi Arabia* 15 (3): 10.

———. 1995. Aiding the Needy Across the World. *Saudi Arabia* 11 (4): 12–17.

Armstrong, Karen. 1996. *Jerusalem: One City, Three Faiths.* New York: Knopf.

Arrieta, Juan Ignacio. 2000. *Governance Structures Within the Catholic Church.* Montreal: Wilson and Lafleur.

Asad, Muhammad. 1954. *The Road to Mecca.* London: Max Reinhardt.

Banks, Lorenzo Livieres. 1970. Notes Toward Understanding the Socio-Historic Functions of the Apostolic Roman Catholic Church of Paraguay. *Revista Paraguaya de Sociologia* 7 (17): 123–132.

Barabasi, Albert-Laszlo, 2002. *Linked: The New Science of Networks.* Cambridge: Perseus.

Barnes, John A. 1968. Networks and Political Process. In Marc Swartz (ed.), 107–130, *Local-Level Politics.* Chicago: Aldine.

Basbous, Antoine. 2002. *L'Arabie Saoudite en Question: Du Wahhabisme á Bin Laden, aux Origines de la Tourmente.* Paris: Perrin.

Baum, Gregory. 1990. Le Rapport Dumont: Démocratiser l'Eglise Catholique. *Sociologie et Sociétés* 22 (2): 115–126.

Bennett, Tony. 2003. Culture and Governmentality. In Jack Z. Bratich, Jeremy Packer, and Cameron McCarthy (eds.), *Foucault, Cultural Studies, and Governmentality.* Albany: State University of New York Press, pp. 47–63.

Bernhart, Joseph. 1939. *The Vatican as a World Power.* London: Longmans and Green.

Bernstein, Saul. 1997. *The Orthodox Union Story: A Centenary Portrayal.* Northvale, NJ: Jason Aronson.

Bertrand, Jean-Rene. 1999. Religions et Territoires. In Jean-René Bertrand and Collette Muller (eds.), *Religions Et Territoires.* Paris: L'Harmattan, pp. 11–16.

Beyer, Peter. 2006. *Religions in Global Society*. New York: Routledge.
Bhatti, Mukhtar. 1981. *Lahore to Mecca by Road*. Lahore: Bhatti Publications.
Biaudet, Henri. 1910. *Les Nonciatures Apostoliques Permanentes Jusqu'en 1648*. Helsinki: Suomalainen Tiedeakatemia.
bin Baz, 'Abd al-'Aziz bin Abdullah. 1990. *A Guide to Hajj, Umrah and Visiting the Prophet's Mosque*. Riyadh: Ifta and Propagation.
Birks, J. S. 1978. *Across the Savannas to Mecca: The Overland Pilgrimage Route from West Africa*. London: C. Hurst.
Blancarte, Roberto. 1998. Religion, Mass Media, and Power. *Religioni e Societa*, 13 (30): 22–34.
Blet, Pierre. 1990. *Histoire de la Représentation Diplomatique du Saint Siège: Des Origines à l'Aube du XIXè siècle*. Vatican City: Archivo Vaticano.
Bogary, Hamza. 1991. *The Sheltered Quarter: A Tale of a Boyhood in Mecca*. Austin: Center for Middle Eastern Studies, University of Texas at Austin.
Bogason, Peter and Juliet A. Musso. 2006. The Democratic Prospects of Network Governance. *American Review of Public Administration*, 36 (1): 3–18.
Bovaird, Tony. 2005. Public Governance: Balancing Stakeholder Power in a Network Society. *International Review of Administrative Sciences*, 71 (2): 217–228.
Boyd, Douglas A. 1970–1971. Saudi Arabian Television. *Journal of Broadcasting* 15 (1): 73–78.
Brasher, B. 2001. *Give Me That Online Religion*. San Francisco: Jossey-Bass.
Bratich, Jack Z., Jeremy Packer, and Cameron McCarthy. 2003. *Foucault, Cultural Studies, and Governmentality*. Albany: State University of New York Press.
Bunt, Gary. 2000. *Virtually Islamic: Computer-Mediated Communication and Cyber Islamic Environments*. Cardiff: University of Wales Press.
Bratich, Graham, Colin Gordon, and Peter Miller (eds.). 1991. *The Foucault Effect: Studies in Governmentality*. Chicago; University of Chicago Press.
Burns, Gene. 1992. *The Frontiers of Catholicism: The Politics of Ideology in a Liberal World*. Berkeley: University of California Press.
Burt, Ronald S. 1980. Models of Network Structure. *Annual Review of Sociology* 6: 79–141.
Burton, Sir Richard F. 1964. *Personal Narrative of a Pilgrimage to Al-Madinah and Meccah*. 2 vols. New York: Dover.
Busse, Heribert. 1968. The Sanctity of Jerusalem in Islam. *Judaism* 17: 441–468.
Buzelais, Alain. 1996. De pecunia vaticani: Ou en sont les finances du Vatican. *Reperes: Bulletin Economique et Financier de la Banque Internationale a Luxembourg* (BIL) 47.
Byrnes, Timothy A. 2001. *Transnational Catholicism in Postcommunist Europe*. Lanham, MD: Rowman and Littlefield.
Calabrese, Omar. 2005. Religious Communication: A Little Structural Anthropology is Necessary. *Religioni e Societa* 20 (53): 63–67.
Campbell, H. 2003. A Review of Religious Computer-Mediated Communication Research. In S. Marriage and J. Mitchell (eds.), *Mediating Religion: Conversations in Media, Culture, and Religion*. Edinburgh: T. and T. Clark/Continuum, pp. 213–228.
Campbell, Heidi H., When Religion Meets New Media. New York: Routledge, 2010.
Caparros, Ernest. 1999. *Code De Droit Canonique Bilingue et Annoté*. Montréal: Wilson and Lafleur.
Capio, Ralph. 1996. The Papacy: A Case Study in Organizational Longevity. *Journal of European Studies* 26 (4): 437–451.
Cardinale, Hyginus Eugene. 1976. *The Holy See and the International Order*. Gerrards Cross, England: Colin Smythe.

Casanova, Jose. 1994. *Public Religions in the Modern World*. Chicago: University of Chicago Press.

———. 1996. Global Catholicism and the Politics of Civil Society. *Sociological Inquiry* 66 (3): 356–373.

———. 1997. Globalizing Catholicism and the Return to a "Universal" Church. In Susanne H. Rudolph and James Piscatori (eds.), *Transnational Religion and Fading States*. Boulder: Westview Press, pp. 121–143.

———. 2001. Religion, the New Millennium, and Globalization. *Sociology of Religion* 62 (4): 415–441.

Castellanos, Renee de la Torre. 2002. The Catholic Diocese: A Transversalized Institution. *Journal of Contemporary Religion* 17 (3): 303–316.

Centro Televisivo Vaticano-Index. "Servizi di Informazione." http://www.vatican.va/news_services/television/index_it.htm. last accessed on July 12, 2010.

Chabin, Michele. 1997. U.S. Orthodox Convert Denied Israel Marriage by Chief Rabbi. *Jewish Bulletin of Northern California*. Retrieved 6/22/2010 (http://www.jewishsf.com/content/2-0-/module/displaystory/ story_id/6943/edition_id/131/format/html/displaystory.html)

Clark, Nicola. 2004. The Expanding Vision of Vatican's Holy See. *International Herald Tribune*. Retrieved 6/22/2010 (http://www.iht.com/articles/2004/04/26/wireless26_ ed3_.php)

Clayman, David. 1999. Cooperation and Tensions between American Jewry and Israel Over Selected Problems Confronting European Jewry. In S. Ilan Troen (ed.), *Jewish Centers and Peripheries: Europe Between American and Israel Fifty Years after World War II*. New Brunswick: Transaction Publishers, pp. 355–371.

Colbi, S. P. 1980. The Christian Establishment in Jerusalem. In Joel L. Kraemer (ed.), *Jerusalem: Problems and Prospects*. New York: Praeger.

Commins, David. 2006. *The Wahhabi Mission and Saudi Arabia*. London: I. B. Tauris.

Congar, Yves. 1970. *L'Eglise*. Paris: Editions du Cerf.

Congregation for the Evangelization of Peoples. 2001. *Norms Governing Diocesan Priests Serving Abroad*. Vatican City.

———. 2001. *Instruction on the Sending Abroad and Sojourn of Diocesan Priests from Mission Territories*. Vatican City.

Cook, K. S. and J. M. Whitmeyer, 1992. Two Approaches to Social Structure: Exchange Theory and Network Analysis. *Annual Review of Sociology* 18: 109–127.

Cook, Michael. 1992. On the Origins of Wahhabism. *Journal of the Royal Asiatic Society of Great Britain and Ireland* 2 (2): 191–202.

Cooke, Miriam and Bruce B. Lawrence. 2005. *Muslim Networks from Hajj to HiP HoP*. Chapel Hill: University of North Carolina Press.

Coghlan, David. 1987. "Corporate Strategy in Catholic Religious Orders." *Long Range Planning* 20 (1): 44–51.

Czarniawska, Barbara (ed.). 2006. *Organization Theory: Current Trends and Disciplinary Reflection*. Northampton, MA: Edward Elgar.

Dabashi, Hamid. 1989. *Authority in Islam: From the Rise of Muhammad to the Establishment of the Ummayads*. New Brunswick, NJ. Transaction Books.

Davies, Jonathan. 2000. The Hollowing Out of Local Democracy and the "Fatal Conceit" of Governing Without Government. *British Journal of Politics and International Relations* 2 (3): 414–428.

Dawson, Lorne L. and Douglas E. Cowan (eds.). 2004. *Religion Online: Finding Faith on the Internet*. New York: Routledge.

De Bouteiller, Georges. 1981. *L'Arabie Saoudite: Cité de Dieu, Cité des Affaires, Puissance Internationale*. Paris: Presses Universitaires de France.

de Chardin, Pierre Teilhard. 1961. *La Messe sur le Monde*. Paris: Editions du Seuil.
Della Cava, Ralph. 1992. Vatican Policy, 1978–90: An Updated Overview. *Social Research* 59 (1): 169–199.
———. 1993. Financing the Faith: The Roman Catholicism. *Journal of Church and State* 35 (1): 1101–1122.
———. 1997. Religious Resource Networks: Roman Catholic Philanthropy in Central and East Europe. In Susanne Hoeber Rudolph and James Piscatori (eds.), *Transnational Religion: Fading States*. Colorado: Westview Press, pp.113–211.
DeLong-Bas, Natana. 2004. *Wahhabi Islam: From Revival and Reform to Global Jihad*. Oxford: Oxford University Press.
Demerath. N. J. 2001. *Crossings the Gods: World Religions and Wordly Politics*. New Brunswick, NJ: Rutgers University Press.
Diehl, Paul F. (ed.). 1997. *The Politics of Global Governance: International Organizations in an Interdependent World*. Boulder, CO: Lynne Rienner.
Din, Adu Bakr Siraj ad-. 1967. Pilgrimage to Mecca. *Studies in Comparative Religion* 1: 171–181.
Diotallevi, Luca. 2004. Roman Catholicism and Religious Globalization. *Antonianum* 79 (2): 345–364.
D'Onorio, Joël-Benoît. 1992. *Le pape et le gouvernement de l'Eglise*. Paris: Fleurus Tardy.
Doreian, Patrick and Franz Stokman. 1997. *Evolution of Social Networks*. New York: Gordon and Breach.
Dorogovtsev, S. N. and J. F. F. Mendes. 2003. *Evolution of Networks: From Biological Nets to the Internet and WWW*. Oxford: Oxford University Press.
Dulles, Avery. 1978. *Models of the Church*. New York: Doubleday.
Dunning, John H. and Jean-Louis Mucchielli (eds.). 2002. *Multinational Firms: The Global-Local Dilemma*. London: Routledge.
Ebaugh, Helen Rose and Janet Saltzman Chafetz (eds.). 2000. *Religion and the New Immigrants: Continuities and Adaptations in immigrant Congregations*. Walnut Creek, CA: AltaMira Press.
———. 2002. *Religion Across Borders: Transnational Immigrant Networks*. Walnut Creek, CA: AltaMira Press.
Eckhardt, Carl C. 1937. *The Papacy and World Affairs*. Chicago: University of Chicago Press.
Elazar, Daniel J. "Land, State, and Diaspora in the History of the Jewish Polity." accessed November 15, 2010. http://www.jcpa.org/dje/articles2/land_state_polity.htm.
———. 1989. *People and Polity: The Organizational Dynamics of World Jewry*. Detroit: Wayne State University Press.
Elazar, Daniel J. and Rela Mintz Geffen. 2000. *The Conservative Movement in Judaism: Dilemmas and Opportunities*. New York: State University of New York Press.
Elazar, Daniel J. and Stuart A. Cohen. 1985. *The Jewish Polity: Jewish Political Organization From Biblical Times to the Present*. Bloomington: Indiana University Press.
Eliade, Mircea (ed.). 1987. *The Encyclopedia of Religion*, vol. 11. New York: Macmillan.
Encel, Sol and Leslie Stein. 2003. *Continuity, Commitment, and Survival: Jewish Communities in the Diaspora*. London: Praeger.
Esposito, John. 1988. *Islam: The Straight Path*. Oxford: Oxford University Press.
Esposito, John (ed.). 1995. *The Oxford Encyclopedia of the Modern Islamic World*. Oxford: Oxford University Press.
———. 1983. *Voices of Resurgent Islam*. Oxford: Oxford University Press.

Falk, Richard. 2001. *Religion and Human Global Governance*. New York: Palgrave Macmillan.
Fandy, Mamoun. 1999. *Saudi Arabia and the Politics of Dissent*. New York: Palgrave Macmillan.
Farsy, Fouad al-. 1990. *Modernity and Tradition: The Saudi Equation*. London: Kegan Paul International.
———. 2001. *Custodian of the Two Holy Mosques King Fahd Bin Abdul Aziz*. Channel Islands: Knight Communications.
Ferrari, Silvio. 1984. The Holy See and the Postwar Palestine Issue: The Internationalization of Jerusalem and the Protection of the Holy Places. *International Affairs* 60 (2): 261–283.
Ferziger, Adam S. 2003. *Training American Orthodox Rabbis to Play a Role in Confronting Assimilation*. Ramat Gan, Israel: Bar Ilan University, the Faculty of Jewish Studies.
Finke, Roger and Patricia Wittberg. 2000. Organizational Revival from Within: Explaining Revivalism and Reform in the Roman Catholic Church. *Journal for the Scientific Study of Religion* 39 (2): 154–170.
Forsgren, Mats and Jan Johanson (eds.). 1992. *Managing Networks in International Business*. Philadelphia: Gordon and Breach.
Foucault, Michel. 1980. *Power/Knowledge: Selected Interviews and Other Writings, 1972–1977*. Edited by Colin Gordon. Translated by Colin Gordon, Leo Marshall, John Mepham, and Kate Soper. Brighton: Harvester.
———. 1991. Governmentality. In Graham Burchell, Colin Gordon, and Peter Miller (eds.), *The Foucault Effect: Studies in Governmentality*. Chicago: University of Chicago Press, pp. 87–104.
Fraser, Cary. 1997. In Defense of Allah's Realm: Religion and Statecraft in Saudi Foreign Policy Strategy. In Susanne H. Rudolph, and James Piscatori (eds.), *Transnational Religion and Fading States*. Boulder, Colorado: Westview Press.
Fredericks, James. 2003. The Catholic Church and the Other Religious Paths: Rejecting Nothing that is True and Holy. *Theological Studies* 64 (2): 225–254.
Friedman, Menachem. 1991. *The Haredi (Ultra-Orthodox) Society: Sources, Trends and Processes* [in Hebrew]. Jerusalem: Jerusalem Institute for Israel Studies.
Fulk, Janet. 2001. Global Network Organizations: Emergence and Future Prospects. *Human Relations* 53 (1): 91–99.
Gal, Allon. 1996. *Envisioning Israel: The Changing Ideals and Images of North American Jews*. Jerusalem: The Magnes Press, and Detroit: Wayne State University Press.
Gal, Allon and Alfred Gottschalk (eds.). 2000. *Beyond Survival and Philanthropy: American Jewry and Israel*. Cincinnati: Hebrew Union College.
Garrett, William. 1992. Canon Law and Corporate Church Action. *Virginia Review of Sociology* 1: 137–148.
Geller, Victor B. 2003. *Orthodoxy Awakens: The Belkin Era and Yeshiva University*. Jerusalem: Urim Publications.
Gellouz, Ezzedine. 1979. *Mecca: The Muslim Pilgrimage*. New York: Paddington Press.
Gerstenfeld, Manfred and Avraham Wyler. 2006. Technology and Jewish Life. *Jewish Political Studies Review* 18 (1–2): 1–22.
Ghoshal, Sumantra and Christopher Bartlett. 1990. The Multinational Corporation as an Interorganizational Network. *Academy of Management Review* 15 (4): 603–625.
Gilliat-Ray, Sophie. 2002. Invitation and Enculturation: The Dissemination and Development of Islam in Contemporary Societies. In John Wolffe (ed.), *Global Religious Movements in Regional Context*. Aldershot, UK: Ashgate, pp. 157–194.

Gnyawali, Devi and Ravindranath Madhavan. 2001. Cooperative Networks and Competitive Dynamics: A Structural Embeddedness Perspective. *Academy of Management Review* 26 (3): 431–445.
Goitein, Solomon D. 1966. The Sanctity of Jerusalem and Palestine in Early Islam. *Studies in Islamic History and Institutions*. Leiden: Brill, pp. 135–148.
Goldberg, Harvey E. 2006. Thinking Globally About Judaism. In Mark Juergensmeyer (ed.), *The Oxford Handbook of Global Religions*. Oxford: Oxford University Press, pp. 175–185.
Goldstein, Israel. 1977. The Role of American Rabbis in World Jewish Affairs, Zionism and the State of Israel. In Gilbert S. Rosenthal (ed.), *The American Rabbi*. New York: Ktav Publishing House.
Goodstein, Laurie. 2008. Pope Adds Meetings with Jewish Leaders, Including Synagogue Visit, to U.S. Itinerary. *New York Times*, April 4, A12.
Gossling, Tobias. 2007. Inside Relationships: A Review of Institutional Approaches towards Multi-Organizational Partnerships, Alliances and Networks. In Tobias Gossling, Leon Oerlemans, and Rob Jansen (eds.), *Inside Networks: A Process View on Multi-Organizational Partnerships, Alliances and Networks* Northampton, MA: Edward Elgar, pp. 13–44.
Gossling, Tobias, Leon Oerlemans, and Rob Jansen (eds.). 2007. *Inside Networks: A Process View on Multi-Organizational Partnerships, Alliances and Networks*. Northampton, MA: Edward Elgar.
Greeley, Andrew. 1998. The Revolutionary Event of Vatican II: How Everything Changed. *Commonwealth* 125 (15): 14–20.
Haddad, Yvonne Yazbeck. 1999. The Globalization of Islam. In John L. Esposito (ed.), *The Oxford History of Islam*. Oxford: Oxford University Press, pp. 601–641.
Hadden, J.K. and D.E. Cowan (eds.). 2000. *Religion on the Internet: Research Prospects and Promises*. Greenwich, CT: JAI Press.
Hajj Management Information System, Ministry of Religious Affairs, Bangladesh. http://www.bdhajjinfo.org/index.php 403.
Hamm, Steve. 2006. O Click All Ye Faithful: The Nun Who Launched the Vatican's Web Site is at Work on a MySpace for Catholics. *Business Week*, May 8.
Hanson, Eric O. 1987. *The Catholic Church in World Politics*. Princeton: Princeton University Press.
Harary, Frank. 1959. Graph Theoretic Methods in the Management Sciences. *Management Science* 6 (4): 387–403.
Harrington, Rachel M. 1997. The Applicability of the Principle of Subsidiarity According to the Code of Canon Law. Ph.D. Dissertation, University of Ottawa, Canada.
Haynes, Jeff. 2001. Transnational Religious Actors and International Politics. *Third World Quarterly* 22 (2): 143–158.
Heanni, Patrick. 2005. *L'Islam de marché: L'Autre revolution conservatrice*. Paris: Seuil.
Hebblethwaite, Peter. 1986. *In the Vatican*. Bethesda, MD: Adler and Adler.
Hefferan, T. 2007. *Twinning Faith and Development: Catholic Parish Partnering in the US and Haiti*. Bloomfield, CT: Kumarian Press.
Helms, Christine Moss. 1981. *The Cohesion of Saudi Arabia: Evolution of Political Identity*. Baltimore: The Johns Hopkins University Press.
Hewson, Martin and Timothy J. Sinclair (eds.). 1999. *Global Governance Theory*. Albany: State University of New York Press.
Hexham, Irving and Karla Poewe. 1997. *New Religions as Global Cultures*. Boulder, CO: Westview Press.
Hilaire, Y. M. 1967. Note sur le recrutement des éléves du Séminaire Français de Rome (1853–1914). *Archives de Sociologie des Religions* 12 (23): 135–140.

Himmelfarb, Harold and Sergio Della Pergola. 1989. *Jewish Education Worldwide: Crosscultural Perspectives*. Maryland: University Press of America.

Hirashima, Hussein Yoshi. 1972. *The Road to Holy Mecca*. Tokyo: Kodansha International.

Hirsch, Richard G. 2000. *From the Hill to the Mount*. Jerusalem: Gefen Publishing House.

Hite, Julie, M. and William S. Hesterly. 2001. The Evolution of Firm Networks: From Emergence to Early Growth of the Firm. *Strategic Management Journal* 22 (3): 275–286.

Hiltz, Starr Roxanne and Murray Turoff. 1993. *The Network Nation*. Cambridge: MIT Press.

Hoffman, Paul. 1984. *O Vatican! A Slightly Wicked View of the Holy See*. New York: Congdon and Weld.

Hoge, Dean R. 1987. *The Future of Catholic Leadership: Responses to the Priest Shortage*. Kansas City, MO: Sheed and Ward.

Hoge, Dean R. and Aniedi Okure. 2006. *International Priests in America: Challenges and Opportunities*. Collegeville, MN: Liturgical Press.

Holliday, Ian. 2000. Is the British State Hollowing-Out? *Political Quarterly* 71 (2): 167–176.

Hopkins, Dwight N., Lois Ann Lorentzen, Eduardo Mendieta, and David Bastone (eds.). 2001. *Religions/Globalization: Theories and Cases*. Durham, NC: Duke University Press.

Imperial, Mark. 2005. Using Collaboration as a Governance Strategy: Lessons from Six Watershed Management Programs. *Administration and Society* 37 (3): 281–320.

Information Office of the Royal Embassy of Saudi Arabia. 1995. Prince Bandar Inaugurates Islamic Center in Ohio. Washington, D.C.: Royal Embassy of Saudi Arabia. Retrieved 6/22/2010 (http://www.saudiembassy.net/archive/1995/news/page17.aspx)

———. 1996. Islamic Projects Sponsored by Saudi Arabia. Washington, D.C.: Royal Embassy of Saudi Arabia. Retrieved 6/22/2010 (http://www.saudiembassy.net/archive/1996/news/page100.aspx)

———. 2000a. Report on Islamic Centers Worldwide. Washington, D.C.: Royal Embassy of Saudi Arabia. Retrieved 11/11/2008 (http://saudiembassy.net/2000News/News/Isldetail.asp?cIndex=2238)

———. 2000b. Islamic University of Medinah: Shining Example of Islamic Education. *Saudi Arabia: The Magazine* 16 (4). Retrieved 6/22/2010 (http://www.saudiembassy.net/files/PDF/Publications/Magazine/2000-winter/Madinah.htm)

———. 2000c. Education Ministry to Send Teachers Overseas to Work. Washington, D.C.: Royal Embassy of Saudi Arabia. Retrieved 6/22/2010 (http://www.saudiembassy.net/archive/2000/news/page290.aspx)

———. 2001a. Hajj Minister Talks about New Umrah Rules at. Washington, D.C.: Royal Embassy of Saudi Arabia. Retrieved 6/22/2010 (http://www.saudiembassy.net/archive/2001/news/page307.aspx)

———. 2001b. New Internet Site on Hajj to be Launched Saturday. Washington, D.C.: Royal Embassy of Saudi Arabia. Retrieved 6/22/2010 (http://www.saudiembassy.net/2001/archive/news/page522.aspx)

———. 2002a. New WAMY Secretary-General Appointed. Washington, D.C.: Royal Embassy of Saudi Arabia. Retrieved 6/22/2010 (http://208.246.28.149/2002News/News/NewsDetail.asp?cIndex=519)

———. 2002b. Information issued for upcoming 1422 Hajj. Washington, D.C.: Royal Embassy of Saudi Arabia. Retrieved 6/22/2010 (http://208.246.28.149/2002/News/News/NewsDetail.asp?cIndex=565)

———. 2002c. King Fahd to Host 1,500 Pilgrims from Asia and Europe. Washington, D.C.: Royal Embassy of Saudi Arabia. Retrieved (http://saudiembassy.net/2000News/News/Hajdetail.asp?cIndex=606)

———. 2002d. Hajj 1420 Sacrificial Meat Being Distributed. Washington, D.C.: Royal Embassy of Saudi Arabia. Retrieved 11/11/2008 (http://saudiembassy.net/2000News/News/Hajdetail.asp?cIndex=626)

———. 2003. King Fahd Hosts More Pilgrims. Washington, D.C.: Royal Embassy of Saudi Arabia. Retrieved 6/22/ 2010 (http://www.saudiembassy.net/archive/2003/news/page551.aspx)

———. 2004. Minister Gives Assurances of Health Preparations for 1424 Hajj. Washington, D.C.: Royal Embassy of Saudi Arabia. Retrieved 6/22/2010 (http://www.saudiembassy.net/archive/2004/news/page796.aspx)

———. 2005a. Saudi Arabian Airline to Transport 100,000 Iranian Pilgrims this Hajj Season. Washington, D.C.: Royal Embassy of Saudi Arabia. Retrieved 6/22/2010 (http://www.saudiembassy.net/archive/2005/news/page877.aspx)

———. 2005b. Over 1.2 Million Hajj Visas Issued. Washington, D.C.: Royal Embassy of Saudi Arabia. Retrieved 6/22/2010 (http://saudiembassy.net/archive/2005/news/page859.aspx)

———. 2006a. Visas for 1426 Hajj. Washington, D.C.: Royal Embassy of Saudi Arabia. Retrieved (http://www.saudiembassy.net/Travel/Hajj.asp)

———. 2006b. Saudi Schools Overseas. Washington, D.C.: Royal Embassy of Saudi Arabia. Retrieved 6/22/2010 (http://www.saudiembassy.net/about/country-information/education/Saudi_schools_overseas.aspx) (http://saudiembassy.net/country/education/edudetail2.asp)

International Telephone and Telegraph Corporation. 1931. *International Communications Review* 7 (1).

Jabbra, Joseph G. and O.P. Dwivedi. (eds.). 2005. *Administrative Culture in a Global Context.* Whitby, Ontario: de Sitter Publications.

Jacobson, Harold K. 1984. *Networks of Interdependence: International Organizations and the Global Political System.* New York: Alfred A. Knopf.

Jessop, Bob. 2004. Multi-level Governance and Multi-level Metagovernance. In Ian Bache and Matthew Flinders (eds.), *Multi-level Governance.* Oxford: Oxford University Press.

John Paul II. 1980. *Norms for the Distribution of Priests.* Vatican City.

———. 1990. *Redemptoris Missio* (encyclical promulgation). Vatican City.

———. 2002. Internet: A New Forum for Proclaiming the Gospel. Message of the Holy Father for the 36th World Communications Day. Pontificium Consilium De Communicationibus Socialibus, May 12. Vatican City.

Jones, Candace, Willian S. Hesterly, and Stephen P. Borgatti. 1997. A General Theory of Network Governance: Exchange Conditions and Social Mechanisms. *Academy of Management Review* 22 (4): 911–945.

Juergensmeyer, Mark. 2003. Thinking Globally About Religion. In Mark Juergensmeyer (ed.), *Global Religions: An Introduction.* New York: Oxford University Press, pp. 3–13.

———. 2006. *The Oxford Handbook of Global Religions.* New York: Oxford University Press.

Kaidi, Hamza. 1980. *Mecca and Medinah Today.* Paris: Les Editions J.A.

Kamal, Ahmad. 1961. *The Sacred Journey: Being Pilgrimage to Makkah.* New York: Duell, Sloan, and Pearce.

Kamber, Michael. 2000. Ban the Web? Not Lubavitch Jews. *Wired*, January 19.

Kechichian, Joseph A. 1986. The Role of the Ulama in the Politics of an Islamic State: The Case of Saudi Arabia. *International Journal of Middle Eastern Studies* 18: 53–71.

Kettl, Donald. 2002. *The Transformation of Governance*. Baltimore: The Johns Hopkins University Press.
Khalifa, Saida Miller. 1977. *The Fifth Pillar: The Story of a Pilgrimage to Mecca and Medina*. Hicksville, NY: Exposition Press.
Khan, Afzal. 2005. American Muslims Perform Hajj, Celebrate Eid-ul Adha. Washington, D.C.: US Department of State. Retrieved 6/22/2010 (http://www.america.gov/st/washfile-english/2005/January/20050121173803cpataruk0.6693384.html)
Khan, Muhammad Zafrulla. 1968. *Pilgrimage to the House of Allah*. London: London Mosque.
Khan, Z. I. and Y. Zaki (eds.). 1986. *Hajj in Focus*. London: Open Press.
Kickert, Walter J., Eric-Hans Klinj, and Joop F. M. Koppenjan (eds.). 1997. *Managing Complex Networks: Strategies for the Public Sector*. Thousand Oaks, CA: Sage.
Kim, Sangmoon and Eui-Hang Shin. 2002. A Longitudinal Analysis of Globalization and Regionalization in International Trade: A Social Network Approach. *Social Forces* 81 (2): 445–471.
King, Henry T., Jr. and Marc A. Le Forestier. 1997. Papal Arbitration. *Dispute Resolution Journal* 52 (3): 74–80.
King, Russel. 1972. The Pilgrimage to Mecca: Some Geographical and Historical Aspects. *Erdkunde* 26: 61–73.
Knox, Hannah, Mike Savage, and Penny Harvey. 2006. Social Networks and the Study of Relations: Networks as Method, Metaphor and Form. *Economy and Society* 35 (1): 113–140.
Knott, Kim. 2005. *The Location of Religion: A Spatial Analysis*. London: Equinox, pp. 110–117.
Knut, Walf. 1979. The Church Tax as a Means of Subsistence. In William Bassett and Petrus Huizing (eds.), *The Finances of the Church*. New York: Seabury Press, pp. 20–27.
Kogut, Bruce and Gordon Walker. 2001. The Small World of Germany and the Durability of National Networks. *American Sociological Review* 66: 317–335.
Komoroczy, Geza. 1999. Jewish Hungary Today: The Jewish Cultural Heritage in the Contemporary Culture of Hungary. In S. Ilan Troen (ed.), *Jewish Centers and Peripheries: Europe Between American and Israel Fifty years After World War II*. New Brunswick: Transaction Publishers, pp. 127–155.
Konhonen, Heli, Reijo Luostarinen, and Lawrence Welch. 1996. Internationalization of SMEs: Inward-Outward Patterns and Government Policy. *Management International Review* 36: 315–329.
Kooiman, Jan. 2003. *Governing as Governance*. London: Sage.
Kowalewski, Mark R. 1993. Firmness and Accommodation: Impression Management in Institutional Roman Catholicism. *Sociology of Religion* 54 (2): 207–217.
La Civita, Michael J. L. 2003. After the Storm. *ONE Magazine* 29, no. 3 (May).
Laguerre, Michel S. 1994. *The Informal City*. Basingstoke: Macmillan Press.
———. 2008. *Global Neighborhoods: Jewish Quarters in Paris, London and Berlin*. Albany: State University of New York Press.
———. 2009. The Transglobal Network Nation: Diaspora, Homeland and Hostland. In Eliezer Ben-Rafael and Yitzhak Sternberg (eds.), *Transnationalism: Diasporas and the Advent of a New (Dis)Order*. Boston: Brill, pp. 195–210.
Laoust, Henri. 1971. Ibn Abd al-Wahhab, Muhammad. In B. Lewis, V. C. Menage, C. Pellat, and J. Schacht (eds.), *The Encyclopedia of Islam*, vol. 3. Leiden: Brill.
Larner, Wendy and William Walters. 2004. *Global Governmentality: Governing International Spaces*. New York: Routledge.
Latham, Robert and Saskia Sassen (eds.). 2005. *Digital Formations: IT and New Architectures in the Global Realm*. Princeton: Princeton University Press.

Laumann, Edward O. and David Knoke. 1986. Social Network Theory. In Siegwart Lindenberg, James S. Coleman, and Stefan Novack (eds.), *Approaches to Social Theory*. New York: Russell Sage Foundation, pp. 83–104.

Lawrence, Bruce B. 2002. Allah On-Line: The Practice of Global Islam in the Information Age. In Stewart M. Hoover and Lynn Schofield Clark (eds.), *Practicing Religion in the Age of the Media: Explorations in Media, Religion, and Culture*. New York: Columbia University Press, pp. 236–253.

Lehmann, David. 2002. Religion and Globalization. In Lina Woodhead et al. (ed.), *Religion in the Modern World*. New York: Routledge, pp. 299–315.

Lemieux, Peter. 2003. A Flicker of Candlelight amid the Darkness. *ONE Magazine* 25, no. 5 (September).

Lernoux, Penny. 1989. *People of God: The Struggle for World Catholicism*. New York: Viking.

Levitt, Peggy. 2007. *God Needs No Passport: Immigrants and the Changing American Religious Landscape*. New York: The New Press.

Lewins, Frank. 1977. Continuity and Change in a Religious Organization: Some Aspects of the Australian Catholic Church. *Journal for the Scientific Study of Religion* 16 (4): 371–382.

Lewis, Bernard. 2003. *The Crisis of Islam: Holy War and Unholy Terror*. New York: Modern Library.

Long, David E. 1997. *The Kingdom of Saudi Arabia*. Gainesville: University Press of Florida.

———. 1979. *The Hajj Today: A Survey of the Contemporary Makkah Pilgrimage*. Albany: State University of New York Press.

Makky, Ghazy Abdul Wahed. 1978. *Mecca: The Pilgrimage City*. London: Croom Helm.

Mani, Muhammad Abdullah al- and Abd ur-Rahman Sbit as-Sbit. 1981. *Cultural Policy in the Kingdom of Saudi Arabia*. Paris: The UNESCO Press.

Mandaville, Peter. 2001. *Transnational Muslim Politics: Reimagining The Umma*. New York: Routledge.

McKern, Bruce. 2003. *Managing the Global Network Corporation*. New York: Routledge.

Mead, Richard. 1998. *International Management: Cross-Cultural Dimensions*. Malden, MA: Blackwell.

Middle East Media Research Institute, University of Pennsylvania. 1985. "Saudi Arabia." In *MEMRI Report*. London: Croom Helm.

Miles, Raymond E., Charles C. Snow, John A. Mathews, Grant Miles, and Henry J. Coleman, Jr. 1997. Organizing in the Knowledge Age: Anticipating the Cellular Form. *Academy of Management Executive* 11 (4): 7–20.

Milward, H. Brinton and Jorg Raab. 2002. Dark Networks: The Structure, Operation, and Performance of International Drug, Terror, and Arms Trafficking Networks. Paper presented at the International Conference on the Empirical Study of Governance, Management, and Performance. Barcelona, Spain, October 4–5.

Mintz, Jerome R. 1992. *Hasidic People: A Place in the New World*. Cambridge: Harvard University Press.

Mitchell, J. Clyde (ed.). 1969. *Social Networks in Urban Situations*. Manchester: Manchester University Press.

Mittelberg, David. 1999. *The Israel Connection and American Jews*. London: Praeger.

Mockler, Robert J. and Dorothy G. Dologite. 1997. *Multinational Cross-Cultural Management: An Integrative Context-Specific Process*. Westport: Quorum Books.

Monge, Peter, and Noshir Contractor. 2003. *Theories of Communication Networks*. New York: Oxford University Press.

———. 2001. Emergence of Communication Networks. In Fredric M. Jablin and Linda L. Putnam (eds.), *The New Handbook of Organizational Communication*. Thousand Oaks, CA: Sage, pp. 440–502.

Montes, Jerome. 2003. Vatican Diplomacy at the Start of the New Millennium. *Futuribles* 284: 45–53.

Moore, Fiona. 2005. *Transnational Business Cultures: Life and Work in a Multinational Corporation*. London: Ashgate.

Moran, Robert T., Philip R. Harris, and William G. Stripp. 1993. *Developing the Global Organization: Strategies for Human Resource Professionals*. Houston: Gulf.

Morosini, Piero. 1998. *Managing Cultural Differences: Effective Strategy and Execution Across Cultures in Global Corporate Alliances*. New York: Pergamon.

Moscati, Italo. 1998. A Taboo: Religion and the Mass Media. *Religioni e Societa* 13 (30): 69–77.

Murphy, Francis X. 1974. Vatican Politics: Structure and Function. *World Politics* 26 (4): 542–559.

Narrowe, Morton H. 1999. From Mosaics to Jews: The Rejection of an Unsuccessful Pattern. In S. Ilan Troen (ed.), *Jewish Centers and Peripheries: Europe Between America and Israel Fifty Years After World War II*. New Brunswick: Transaction Publishers, pp. 181–207.

Neuvecelle, Jean. 1955. *The Vatican: Its Organization, Customs, and Way of Life*. New York: Criterion Books.

Newman, Mark, and Albert Laszlo Barabasi (eds.). 2006. *The Structure and Dynamics of Networks*. Princeton: Princeton University Press.

Newman, Janet. 2005. Enter the Transformational Leader: Network Governance and the Micro-politics of Modernization. *Sociology* 39 (4): 717–734.

———.1929 Lateran Treaty. 1930. In John W. Wheeler-Bennett (ed.), *Documents on International Affairs*. London: Oxford University Press, pp. 216–241.

Nye, Joseph and Robert Keohane. 1971. Transnational Relations and World Politics: An Introduction. *International Organization* 25: 329–346.

Oakley, Francis. 2004. Constitutionalism in the Church. In Francis Oakley and Bruce Russett (eds.), *Governance, Accountability, and the Future of the Catholic Church*. New York: Continuum, pp. 76–87.

Oerlemans, Leon, Tobias Gossling, and Rob Jansen. 2007. *Inside Networks: A Process View on Multi-organisational Partnerships, Alliances and Networks*. In Tobias Gossling, Leon Oerlemans, and Rob Jansen (eds.), Inside Networks: A Process View on Multi-organisational Partnerships, Alliances and Networks. Northampton, MA: Edward Elgar, pp. 3–12.

Oliver, Willard M. 2006. 500 Years of Service to the Popes. *Crime and Justice International* 22 (90): 4–10.

Ortiz, Renato. 2003. Notes on Religion and Globalization. *Nepantla: Views from South* 4 (3): 423–448.

Owen, Richard. 2006. Vatican Cashes in by Putting Price on the Pope's Copyright. *The Times* (London), January 23.

Palard, Jacques. 1997. *Le gouvernement de l'Eglise Catholique: Synodes et exercices du pouvoir*. Paris: Cerf.

Pallenberg, Corrado. 1960. *Inside the Vatican*. New York: Hawthorn Books.

Park, Seung Ho. 1996. Managing an Interorganizational Network: A Framework of the Institutional Mechanism for Network Control. *Organization Studies* 17: 795–824.

Parkhe, Arvind, Stanley Wasserman, and David A. Ralston. 2006. New Frontiers in Network Theory Development. *Academy of Management Review* 31 (3), 560–568.

Pearson, Michael. 1996. *Pilgrimage to Mecca: The Indian Experience 1500–1800*. Princeton: Markus Wiener.

Peters, Francis E. 1986. *Jerusalem and Mecca: The Typology of the Holy City in the Near East*. New York: New York University Press.

———. 1994. *The Hajj: The Muslim Pilgrimage to Mecca and the Holy Places*. Princeton: Princeton University Press.

Peters, Guy and John Pierre. 1998. Governance Without Government? Rethinking Public Administration. *Journal of Public Administration Research and Theory* 8 (2): 223–243.

Phatak, Arvind V., Rabi S. Bhagat, and Roger J. Kashlak. 2005. *International Management: Managing in a Diverse and Dynamic Global Environment*. New York: McGraw-Hill.

Podolny, Joel M. and Karen L. Page. 1998. Network Forms of Organization. *Annual Review of Sociology* 24: 57–76.

Pollard, John F. 2005. *Money and the Rise of the Modern Papacy: Financing the Vatican, 1850–1950*. New York: Cambridge University Press.

Pontifical Council for Social Communications. 2002. *The Church and Internet*. Papal Documents. Vatican: The Holy See Website. Retrieved 6/22/2010 (http://www.vatican.va/roman_curia/pontifical_councils/pccs/documents/rc_pc_pccs_doc_20020228_church-internet_en.html)

———. 2002. Ethics in Internet. Papal Documents. Vatican: The Holy See Website. Retrieved 6/22/2010 (http://www.vatican.va/roman_curia/pontifical_councils/pccs/documents/rc_pc_pccs_doc_20020228_ethics-internet_en.html#top)

Paul VI. 1967. Regimini Ecclesiae Universalis. August 1967 decree.

Powell, Walter. 1990. Neither Market nor Hierarchy: Network Forms of Organization. *Research In Organizational Behavior* 12: 295–336.

Quandt, William. 1981. *Saudi Arabia in the 1980s: Foreign Policy, Security, and Oil*. Washington, D.C.: The Brookings Institution.

Rai, Saritha. 2004. Short on Priests: US Catholics Outsource Prayers to Indian Clergy. *New York Times*, June 13.

Rasheed, Madawi al-. 2002. *A History of Saudi Arabia*. New York: Cambridge University Press.

Rasheed, Madawi al- and Robert Vitalis (eds.). 2004. *Counter-Narratives: History, Contemporary Society, and Politics in Saudi Arabia and Yemen*. New York: Palgrave.

Rashid, Nasser Ibrahim and Esber Ibrahim Shaheen. 1995. *Saudi Arabia All You Need to Know*. Joplin, MO: International Institute of Technology.

Reese, Thomas J. 1996. *Inside the Vatican: The Politics and Organization of the Catholic Church*. Cambridge: Harvard University Press.

———. 1992. *A Flock of Shepherds: The National Conference of Catholic Bishops*. Kansas City, MO: Sheed and Ward.

———. 1989. *Archbishop: Inside the Power Structure of the American Catholic Church*. San Francisco: Harper and Row.

Rentz, George Snavely, Jr. 1948. *Muhammad ibn 'Abd al-Wahhab (1703/4–1792) and the Beginnings of the Unitarian Empire in Arabia*. PhD Dissertation, University of California, Berkeley.

Richard, P. 1906. Les Origines des Nonciatures Permanentes. *Revue d'Histoire Ecclésiastique* 7 : 52–70, 317–338.

Rhodes, Rod A. W. 1988. *Beyond Westminster and Whitehall*. London: Unwin Hyman.

Roff, William R. 1982. Sanitation and Security: The Imperial Powers and the Nineteenth Century Hajj. *Arabian Studies* 6: 143–160.

Romero, Catalina. 2001. Globalization, Civil Society and Religion from a Latin American Standpoint. *Sociology of Religion* 62 (4): 475.

Rosenau, James N. and Ernst-Otto Czempiel (eds.). 1992. *Governance Without Government: Order and Change in World Politics*. Cambridge: Cambridge University Press.
Rosenthal, Gilbert S. (ed.). 1977. *The American Rabbi*. New York: Ktav Publishing House.
Rowley, Gwyn and S. A. S. El-Hamdan. 1978. The Pilgrimage to Mecca: An Exploratory and Predictive Model. *Environment and Planning A* 10: 1053–1071.
———. 1977. Once a Year in Mecca. *Geographical Magazine* 49: 753–759.
Rugh, William. 1980. Saudi Mass Media and Society in the Faisal Era. In Willard A. Beling (ed.), *King Faisal and the Modernisation of Saudi Arabia*. London: Croom Helm.
Sack, R. D. 1986. *Human Territoriality, Its Theory and History*. London: Cambridge.
Sambamurthy, Vallabh and Robert W. Zmud. 1999. Arrangements for Information Technology Governance: A Theory of Multiple Contingencies. *MIS Quarterly* 23 (2): 261–290.
Samman, Khaldoun. 2007. *Cities of God and Nationalism*. Boulder, CO: Paradigm.
Sardar, Ziauddin. 1984. The Future of Hajj: Some Basic Considerations. *Islamic Culture* 58 (4): 307–326.
Sassen, Saskia (ed.). 2007. *Deciphering the Global: Its Scales, Spaces and Subjects*. New York: Routledge.
Saudi Arabia Information. 2003. WAMY Builds Mosques in Southern Makkah. Riyadh, Saudi Arabia: Saudinf. Retrieved 6/22/2010 (http://www.saudinf.com/main/y6370.htm).
———. 2004a. Prince Sultan Donates $2 Million for Kazakhstan Mosque. Riyadh, Saudi Arabia: Saudinf. Retrieved 6/22/2010 (http://www.saudinf.com/main/y7544.htm)
———. 2004b. 1,947,578 Copies of Quran Distributed within the Kingdom and Abroad. Riyadh, Saudi Arabia: Saudinf. Retrieved 6/22/2010 (http://www.saudinf.com/main/y7578.htm)
———. 2005. WAMY Builds Mosques. Riyadh, Saudi Arabia: Saudinf. Retrieved 6/22/2010 (http://www.saudinf.com/main/y8450.htm)
Schmool, Marlena. 1999. British Jewry: Prospects and Problems. In S. Llan Troen (ed.), *Jewish Centers and Peripheries: Europe Between American and Israel Fifty Years After World War II*. New Brunswick: Transaction Publishers, pp. 227–252.
Seldes, George. 1934. *The Vatican: Yesterday, Today, Tomorrow*. New York: Harper.
Shair, I. M., and P. P. Karan. 1979. Geography of the Islamic Pilgrimage. *GeoJournal* 3 (6): 599–608.
Sharot, Stephen. 2001. *A Comparative Sociology of World Religions*. New York: New York University Press.
Sivan, Emmanuel. 1967. Le Caractére Sacré de Jérusalem dans l'Islam aux XIIé-XIIIe Siecles. *Studia Islamica* 27: 149–182.
Smith, Huston. 1958. *The Religions of Man*. New York: Harper.
Smith, William L. 2004. *Irish Priests in the US: A Vanishing Subculture*. Dallas, TX: University Press of America.
Snow, Charles C., Raymond E. Miles, and Brent B. Alfred. 2003. Managing Leading-Edge Multinational Corporations. In Bruce McKern (ed.), *Managing the Global Network Corporation*. New York: Routledge, pp. 201–219.
Sorensen, Eva. 2006. Metagovernance: The Changing Role of Politicians in Processes of Democratic Governance. *American Review of Public Administration* 36 (1): 98–114.

Souryal, Sam S. 1987. The Religionization of a Society: The Continuing Application of Shariah Law in Saudi Arabia. *Journal for the Scientific Study of Religion* 26 (4): 429–449.

Stackhouse, Max L. et al. (eds.). 2000. *God and Globalization*. 3 vols. Harrisburg, PA: Trinity Press International.

Stoker, Gerry. 2006. Public Value Management: A New Narrative for Networked Governance? *American Review of Public Administration* 36 (1): 41–57.

Strenski, Ivan. 2004. The Religion in Globalization. *Journal of the American Academy of Religion* 72 (3): 631–652.

Talmon, Shemaryahu. 1974. The Biblical Concept of Jerusalem. In J. M. Oesterreicher and A. Sinai (eds.), *Jerusalem*. New York: John Day.

Thompson, Grahame. 2003. *Between Hierarchies and Markets:The Logic of Network Forms of Organizations*. Oxford: Oxford University Press.

Tichy, Noel M., Michael L. Tushman, and Charles Fombrun. 1979. Social Network Analysis for Organizations. *Academy of Management Review* 4 (4): 507–519.

Treistman, Robin. 2000. Judaic Law in an Internet World. 66th IFLA Council and General Conference. Retrieved 6/22/2010 (http://www.ifla.org/IV/ifla66/papers/020-160e.htm)

Troen, S. Ilan. 1999. Introduction: The Post-Holocaust Dynamics of Jewish Centers and Peripheries. In S. Ilan Troen (ed.), *Jewish Centers and Peripheries: Europe Between American and Israel Fifty Years After World War II*. New Brunswick: Transaction Publishers, pp. 1–26.

Tully, Shawn and Marta F. Dorion. 1987. The Vatican's Finances. *Fortune* 116 (14), December 21.

Turner, Cuthbert Hamilton. 1899–1933. *Ecclesiae Occidentalis Monumenta Juris Antiquissima*. 5 vols. Oxford: Clarendon.

Urbach, Efraim E. 1968. The Heavenly and the Earthly Jerusalem in Rabbinic Thought. In *Jerusalem Through the Ages*. Jerusalem: Israel Exploration Society, pp. 156–171.

Urritia, Francisco Javier. 1985. Delegation of the Executive Power of Governance. *Studia Canonica. Revue Canadienne de Droit Canonique* 19 (1): 339–355.

Uthaymin, Abd Allah al-Salih al-. 1972. Muhammad Ibn 'Abd al-Wahhab: The Man and His Works. Ph.D. dissertation. Edinburgh University.

Vaillancourt, Jean-Guy. 1980. *Papal Power: A Study of Vatican Control over Catholic Lay Groups*. Berkeley: University of California Press.

Vallier, Ivan. 1972. The Roman Catholic Church: A Transnational Actor. In Robert O. Keohane and Joseph Nye (eds.), *Transnational Relations and World Politics*. Cambridge: Harvard University Press, pp. 479–502.

Vasquez, Manuel A. and Marie F. Marquardt. 2003. *Globalizing the Sacred: Religion Across the Americas*. New Brunswick, NJ: Rutgers University Press.

Vassiliev, Alexei. 1998. *The History of Saudi Arabia*. London: Saqi Books.

Vegh, Steven. 2004. Foreign Priests Fill Persistent Vacancies in the US Dioceses. Dozens Have Come to VA from Abroad. *The Virginia Pilot and the Ledger-Star*, January 4, A1.

Vital, David. 1990. *The Future of Jews*. Cambridge, MA: Harvard University Press.

Voll, John. 1994. *Islam: Continuity and Change in the Modern World*. Syracuse, NY: Syracuse University Press.

———. 1975. Muhammad Hayat al-Sindi and Muhammad Ibn 'Abd al-Wahhab: An Analysis of an Intellectual Group in Eighteenth Century Medina. *Bulletin of the School of Oriental and African Studies* 38 (1): 32–39.

Wach, Joachim. 1955. *Types of Religious Experience: Christian and Non-Christian*. London: Routledge and Kegan Paul.

Walsh, Michael J. 1983. *Vatican City State*. New York: Harper.

Warren, Roland L. 1967. The Interorganizational Field as a Focus for Investigation. *Administrative Science Quarterly* 12: 396–419.
Wasserman, Stanley and Katherine Faust. 1994. *Social Network Analysis*. Cambridge: Cambridge University Press.
Watts, Duncan J. 2004. The "New" Science of Networks. *Annual Review of Sociology* 30: 243–270.
Weber, Max. 1948. The Social Psychology of the World Religions. In H. H. Gerth and C. Wright Mills (eds.), *From Max Weber: Essays in Sociology*. London: Routledge and Kegan Paul, pp. 267–301.
Welch, Lawrence S. and ReiJo K. Luostarinen. 1993. Inward-Outward Connections in Internationalization. *Journal of International Marketing* 1 (1): 44–56.
Wellman, B. and S. D. Berkowitz (eds.). 1988. *Social Structures: A Network Approach*. New York: Cambridge University Press.
Wellman, B. and C. Haythornthwaite (eds.). 2002. *The Internet in Everyday Life*. Oxford: Blackwell.
Wells, Alan and Mary Lou De Ciantis. 1985. World Systems: Transnational Organizations and the Catholic Church. *Free Inquiry in Creative Sociology* 13 (1): 70–74.
Werblowsky, R. J. Zwi. 1973–1974. Jerusalem: Holy City of Three Religions. *Jaarbericht ex Oriente Lux* 23: 423–39.
Wertheimer, Jack. 1987. *The American Synagogue: A Sanctuary Transformed*. Cambridge: Cambridge University Press;
Wertheimer, Jack (ed.). 2000. *Jews in the Center: Conservative Synagogues and their Members*. New Brunswick: Rutgers University Press.
West, D. B. 1996. *Introduction to Graph Theory*. Upper Saddle River, NJ: Prentice-Hall.
Wheeler, Brannon. 2006. *Mecca and Eden: Ritual, Relics, and Territory in Islam*. Chicago: University of Chicago Press.
Whitten, Norman E. and Alvin W. Wolfe. 1973. Network Analysis. In John J. Honigmann (ed.), *Handbook of Social and Cultural Anthropology*. Chicago: Rand McNally, pp. 717–746.
Williams, George H. 1982. John Paul II's Concepts of Church, State, and Society. *Journal of State and Church* 24: 463–483.
Williamson, Benedict. 1929. *The Treaty of the Lateran*. London: Burns Oates and Washbourne.
Wimmer, Andreas and Nina Glick Schiller. 2002. Methodological Nationalism, the Social Sciences, and the Study of Migration: An Essay in Historical Epistemology. *Global Networks* 2 (4): 301–334.
Wooden, Cindy. 2009.Pope Asks Young Catholics to Use Technology to Share their Faith. *Catholic News Service*, January 23.
Wynbrandt, James. 2004. *A Brief History of Saudi Arabia*. New York: Facts On File.
Yassini, Ayman al-. 1995. Ibn 'Abd al-Wahhab, Muhammad. In John L. Esposito (ed.), *The Oxford Encyclopedia of the Modern Islamic World*, Vol. 2. New York: Oxford University Press.
———. 1985. *Religion and State in the Kingdom of Saudi Arabia*. Boulder, Colorado: Westview Press.
Young, Katherine K. 1992. World Religions: A Category in the Making? In Michel Despland and Gerard Vallee (eds.), *Religion in History: The Word, the Idea, the Reality*. Waterloo, IA: Wilfrid Laurier University Press, pp. 111–130.
Zaid, Abdulla Mohamed al-. 1982. *Education in Saudi Arabia: A Model With Difference*. Jeddah: Tihama.
Zakat and Ushr. 2006. Hajj Policy—2006. Islamabad, Pakistan: Government of Pakistan, Ministry of Religious Affairs, Retrieved 12/4/2008 (http://www.pakistan.gov.pk/divisions/religious-affairs-division/media/HP2006.pdf.)

Ziauddin, Sardar and M. A. Zaki Badawi. 1977. *Hajj Studies* vol. 1. London: Croom Helm.

Zigarelli, Michael A. 1993. Catholic Social Teaching and the Employment Relationship: A Model for Managing Human Resources in Accordance with Vatican Doctrine. *Journal of Business Ethics* 12 (1): 75–82.

Index

C

D

O

T

For Product Safety Concerns and Information please contact our EU representative GPSR@taylorandfrancis.com
Taylor & Francis Verlag GmbH, Kaufingerstraße 24, 80331 München, Germany

www.ingramcontent.com/pod-product-compliance
Ingram Content Group UK Ltd.
Pitfield, Milton Keynes, MK11 3LW, UK
UKHW022120190926
475702UK00013B/1548

* 9 7 8 0 4 1 5 8 4 9 0 3 6 *